CHINA'S REGIONS IN AN ERA OF GLOBALIZATION

'With considerable analytical rigor and clarity in exposition, Summers convincingly shows that China's engagement with globalization cannot just be understood at the national level, but has fundamentally been a story about differentiated participation in the global economy across China's regions. This is the first book to examine China's post-1978 development from a regional perspective. Students, researchers, and policy makers who want to understand China's rapid economic rise in the 21st century will find this book indispensable.'

Alvin So, The Hong Kong University of Science and Technology, Hong Kong, China

The rise of China has been shaped and driven by its engagement with the global economy during a period of intensified globalization, yet China is a continent-sized economy and society with substantial diversity across its different regions. This means that its engagement with the global economy cannot just be understood at the national level, but requires analysis of the differences in participation in the global economy across China's regions.

This book responds to this challenge by looking at the development of China's regions in this era of globalization. It traces the evolution of regional policy in China and its implications in a global context. Detailed chapters examine the global trajectory of what is now becoming known as the Greater Bay Area in southern China, the globalization of the inland mega-city of Chongqing, and the role of China's regions in the globally-focused belt and road initiative launched by the Chinese government in late 2013.

The book will be of interest to practitioners and scholars engaging with contemporary China's political economy and international relations.

Tim Summers works on the political economy and international relations of contemporary China. He is a Lecturer at the Centre for China Studies, The Chinese University of Hong Kong, and a (non-resident) Senior Consulting Fellow on the Asia Programme at Chatham House. He was British consul-general in Chongqing from 2004 to 2007.

Routledge Contemporary China Series

China Reclaims World Power Status
Putting an End to the World America Made
Paolo Urio

The Economic Roots of the Umbrella Movement in Hong Kong
Globalization and the Rise of China
Louis Augustin-Jean and Anthea H.Y. Cheung

China's Hydro-politics in the Mekong
Conflict and Cooperation in Light of Securitization Theory
Sebastian Biba

Economic Policy Making in China (1949–2016)
The Role of Economists
Pieter Bottelier

The Power of Relationalism in China
Leah Zhu

China's Financial Opening: Coalition Politics and Policy Changes
Yu-Wai Vic Li

Urbanization, Regional Development and Governance in China
Jianfa Shen

Midwifery in China
Ngai Fen Cheung and Rosemary Mander

China's Virtual Monopoly of Rare Earth Elements
Economic, Technological and Strategic Implications
Roland Howanietz

China's Regions in an Era of Globalization
Tim Summers

For more information about this series, please visit: www.routledge.com/Routledge-Contemporary-China-Series/book-series/SE0768

CHINA'S REGIONS IN AN ERA OF GLOBALIZATION

Tim Summers

First published 2018
by Routledge
2 Park Square, Milton Park, Abingdon, Oxon OX14 4RN

and by Routledge
711 Third Avenue, New York, NY 10017

Routledge is an imprint of the Taylor & Francis Group, an informa business

© 2018 Tim Summers

The right of Tim Summers to be identified as author of this work has been asserted by him in accordance with sections 77 and 78 of the Copyright, Designs and Patents Act 1988.

Chatham House, the Royal Institute of International Affairs, is an independent policy institute and does not express opinions of its own. The views expressed in their affiliated books are those of the contributors. These authors may or may not be affiliated directly with Chatham House.

All rights reserved. No part of this book may be reprinted or reproduced or utilised in any form or by any electronic, mechanical, or other means, now known or hereafter invented, including photocopying and recording, or in any information storage or retrieval system, without permission in writing from the publishers.

Trademark notice: Product or corporate names may be trademarks or registered trademarks, and are used only for identification and explanation without intent to infringe.

British Library Cataloguing in Publication Data
A catalogue record for this book is available from the British Library

Library of Congress Cataloging in Publication Data
Names: Summers, Tim, author.
Title: China's regions in an era of globalization / Tim Summers.
Description: Abingdon, Oxon ; New York, NY : Routledge, 2018. | Series: Contemporary China | Includes bibliographical references and index.
Identifiers: LCCN 2018001311| ISBN 9781138682245 (hardback) | ISBN 9781138682252 (pbk.) | ISBN 9781315545295 (ebook)
Subjects: LCSH: China—Foreign economic relations. | Regional policy—China. | Globalization.
Classification: LCC HF1604 .S85 2018 | DDC 337.51—dc23
LC record available at https://lccn.loc.gov/2018001311

ISBN: 978-1-138-68224-5 (hbk)
ISBN: 978-1-138-68225-2 (pbk)
ISBN: 978-1-315-54529-5 (ebk)

Typeset in Bembo
by Florence Production Ltd, Stoodleigh, Devon, UK

To the memory of Arif Dirlik (1940–2017)

CONTENTS

List of illustrations   ix
Acknowledgements and note   xi

    Introduction   1

1 Globalization and China   3

2 Regional policy in contemporary China: a historical overview   15

3 The Greater Bay Area: from factory floor to global challenger   41

4 Chongqing: globalization moves inland   63

5 The belt and road initiative and China's regions   83

6 Conclusion: China and its regions in an era of globalization   105

Bibliography   115
Index   129

ILLUSTRATIONS

Figures

0.1	China, Eurasia, and the belt and road	xiii
2.1	China's local government system	16

Tables

3.1	Pearl River Delta economic and social data (2015)	42
3.2	Pearl River Delta economic targets	46
3.3	Guangdong thirteenth five-year programme, key targets for 2020	50
3.4	Comparison of economic data (2014)	56
4.1	Chongqing economic data	65
5.1	China–Europe freight trains (at March 2017)	94

Box

3.1	Pearl River Delta concepts	42

ACKNOWLEDGEMENTS

As ever in projects of this scale, lots of people have offered support and inspiration, and I will not attempt to list them all here. In particular, colleagues and mentors past and present at The Chinese University of Hong Kong (CUHK), Chatham House, and the Foreign and Commonwealth Office, as well as students at CUHK over the past decade have helped me develop my thinking. I am grateful to Andrew Taylor and colleagues at Taylor and Francis for their interest in the project and for their help in seeing the book through to publication. This book was originally intended as part of a Chatham House series and I am grateful to Caroline Soper, Amanda Moss, and colleagues at Chatham House for their support. Anonymous reviewers of the original outline and the first draft provided insightful comments and suggestions. At home, Lucy and George have been as supportive and inspiring as usual!

Note

Chinese names appear in the order in which they are used in China, surname followed by given name.

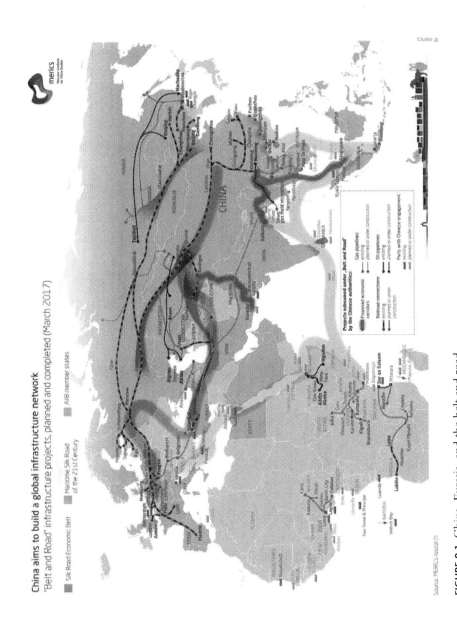

FIGURE 0.1 China, Eurasia, and the belt and road

Source: Mercator Institute for China Studies. Printed with permission.

INTRODUCTION

Since the 1980s, China's rapid economic rise has been accompanied by transformations in the global economy under a new and intensified form of globalization. These two phenomena have been intimately related: the rise of China has been shaped and driven by its incorporation into the global economy, while the nature and structures of globalization have been radically changed as a result of China's development.

This book will argue that China's engagement with globalization over this period cannot just be understood at the national level, but has fundamentally been a story about differentiated participation in the global economy across China's regions. In doing so, the book goes beyond many existing studies to place China's regions more deliberately in a global context.

That there exists a sub-national dimension to China's relationship with globalization is not surprising. China is a continent-sized country, economy, and society with substantial regional diversity. Geographical and historical conditions for external economic interactions vary significantly across China, from the long-standing external ties of south-eastern coastal provinces of Guangdong and Fujian with southeast Asia and beyond, to the inland or western parts of the country that were cut off from the global economy for much of recent history. Levels of economic development also vary substantially across the country. Writing his bullish account of China's future development, Chinese writer Zhang Weiwei suggests that China should be understood as 'a country made up of at least two groups of regions', the 'quasi-developed countries' and the 'emerging economies', and that the 'dynamic and mutually complementary interactions between them' is part of the reason why China has risen so quickly.[1]

There is therefore much to be gained from studying China's regions in this era of globalization. Chapter 1 of this book puts this in context by discussing globalization – in particular its economic dimensions – and China's relationship with it. Chapter 2 explores the development of regional policy and China's regions since the establishment of the People's Republic of China (PRC) in 1949, with a particular focus on where changes in regional policy and development during this period have not only been a function of domestic dynamics, but have also been devised in response to the global context.

The different trajectories of China's regions in this era of globalization are explored in detail from three different perspectives in Chapters 3, 4 and 5. Chapter 3 looks at what is now becoming known as the Greater Bay Area between the southern coastal province of Guangdong and Hong Kong and Macao, the part of China that has been at the forefront of China's engagement with globalization since the 1980s. This stimulated the desire for similar forms of engagement with the global economy across many parts of China, and Chapter 4 looks at the response from Chongqing, a city that grasped this challenge in a proactive and often innovative way. Part of this was the municipality's initiative in developing new railway freight routes across land through central Asia to Europe, a development that has become central to the belt and road initiative, launched by China's national leadership in late 2013. Chapter 5 looks at the role of China's regions in the belt and road.

In conclusion, Chapter 6 discusses the implications of the preceding chapters for understanding China's present and future relationships with globalization. Insights into the changing relationship between China and globalization are all the more important now that China has become one of the most influential global economic powers (however this is measured), especially at a point in time when it appears the current phase of globalization may be giving way to something else. The answers to current debates about whether this is 'de-globalization' or a globalization that looks different in form and substance will depend at least partly on the approach of China's regions over the coming years.

The book does not claim to be an exhaustive treatment of all the global interactions of China's regions. Its focus is on the political economy of China's regions and the policy developments that shed light on their relationship to globalization, in the process contributing to our understanding of the history of the 'reform era' in China. Research has been carried out through textual analysis and interpretation of policy documents and media reports, discussions with individuals and participation in events over many years of work and travel across China, supplemented by relevant secondary sources. The choice of regions for Chapters 3 and 4 – the Pearl River Delta and Chongqing – partly reflects the author's own experience of working in China, while the analysis of the belt and road initiative flows naturally from previous work on Yunnan.[2]

Notes

1. Zhang Weiwei. *The China wave: rise of a civilizational state*. Singapore: World Scientific, 2011, p. 33.
2. Tim Summers. *Yunnan: a Chinese Bridgehead to Asia: a case study of China's political and economic relations with its neighbours*. Oxford, UK: Chandos, 2013.

1
GLOBALIZATION AND CHINA

The context for this exploration of China's regions in an era of globalization begins with a discussion of what is meant by 'globalization' itself. This is not such a straightforward question to answer, as even though globalization may have become a 'paradigmatic' concept in describing global political economy since the 1980s,[1] it remains a highly contested one. Interpretations of globalization have ranged from seeing it as 'neutral' economic and commercial processes driven by growing trade and investment to the globalization of capitalism following the end of the Cold War, or as the political and cultural spread of 'Western', especially US, influence.[2] A conventional view of globalization sees it as 'an objective process of market integration on a global scale, driven by developments in transportation and communication technologies'.[3] This economic emphasis is at the heart of many understandings of globalization – including, as we will see, in China – and the role of business in globalization is a key part of its emergence. But globalization is about more than economic integration. The changes in markets, trade and investment, and business operation are intertwined with politics and social issues, from questions of where power lies in the global economy to the much-debated problems of socio-economic inequality between and within countries.[4]

As many writers have pointed out,[5] globalization is not a new phenomenon, and it would be more accurate to refer to a 'recent' or 'current' phase of globalization when talking about developments from the 1980s onwards. The earlier phase of globalization this period followed had been characterized by the Bretton Woods institutions (World Bank, International Monetary Fund) which were key to the management of parts of the global economy – at that stage, this did not include China – in the decades after World War II. The features of this period included growth in international trade, but also a reasonably strong role for national governments in the management of flows of goods and capital across their borders.

However, this balance did not survive the economic shocks of the 1970s, and in the 1980s and 1990s the Bretton Woods monetary regime was superseded by 'a more ambitious agenda of economic liberalization and deep integration', what Harvard economist Dani Rodrik calls 'hyperglobalization'.[6] The choice of terms to describe this has reflected intense academic debates about how to characterize and explain this period.[7]

The main characteristics of this period range from the emerging dominance of multi-national corporations across many sectors of the economy, the changing nature of transborder trade and investment as well as its rapid growth, to the growth of financial globalization, and the role of national governments and global institutions in facilitating these developments. The ways that these developments are interlinked are complex, but they have taken place in the context of the spread of capitalism, and an ideology of 'free enterprise as the key to continuous economic progress'.[8] The spread of this capitalist globalization in the 1980s was partly enabled by the 'opening up' of China (as we will discuss further below), and subsequently by the end of the Cold War.

A second important feature of this phase of globalization of particular relevance to understanding China's relationship with globalization is the changing nature of production. During this period a 'business revolution' has stimulated the transformation of national firms into transnational (or multinational) corporations through the development of supply chains and outsourcing on a global scale.[9] These global production networks constitute an 'international division of labour, in which each function or discrete stage of a value chain is spatially or geographically relocated in the most efficient site, and undertaken by different firms'.[10] Such production networks were not a novelty of the 1980s, and in east Asia had been developed with Japanese corporations in the lead from the 1960s onwards; but the scale and pace of their development on a global scale grew substantially under the new phase of globalization from the 1980s onwards, spurred by technological development, which enabled the rapid exchange of information on a global scale.

These phenomena have not just been driven by corporations. National governments and international institutions such as the World Bank and IMF have facilitated these developments through measures to liberalize international trade and investment, as have domestic policies that have cemented the rights of foreign investors, increased labour flexibility, and reduced corporate taxes. In doing this, relationships between states have often become competitive as governments have sought to maximize national competitiveness in attracting investment, leading to convergence across countries in many areas of economic policy. The outcomes have included increased power of capital relative to labour, a somewhat circumscribed space for state policies, and state capture by business interests, what some have described as 'elected plutocracy'. The idea that states have been passive responders to the influence of global capital has also stimulated debate over the extent to which globalization has challenged the pre-eminence of the nation-state as a 'power container', as the transformation of states and their economic interdependence during this period of globalization has had an impact on international relations by diluting the ability of national governments to set the policy agenda or control outcomes. This has also enhanced the relevance of looking at the involvement of sub-national governments in international affairs (sometimes known as 'paradiplomacy').[11]

The resulting structures of global political economy have enabled the industrial organization of production on a global basis, with the land, labour and resources available in 'new frontiers' (such as China from the 1980s) being pulled into more complex global production hierarchies; this has been further stimulated after the demise of the socialist alternatives to capitalist modernity and decline of anti-systemic movements, which previously provided the focus for labour resistance.[12] Production has become fragmented across multiple geographies and firms, producing new organizational structures in the corporate world. In many sectors, though, in spite of this fragmentation, supply chains have been organized and managed by large oligopolistic 'system integrator' firms – from Apple to Boeing – which are able to play this role because they have access to the capital and human resources to engage in the intensive research and development activities necessary to remain globally competitive.[13] Potential new entrants, particularly from developing economies, are consequently required to climb a steep and competitive curve in order to stand a chance of becoming significant players on a global – and sometimes even a national – playing field.

For all the talk of market economies and competition, globalization has therefore not led to a 'flat world'. The intensified hierarchies and the dominance in many sectors of a small number of firms means that the 'commanding heights' of much of the global economy have – at least until very recently[14] – remained predominantly in the hands of firms based in developed economies, especially the US, Europe, and Japan. This raises specific challenges for firms and governments in emerging economies such as China, and this is an important factor informing industrial policy in China today.

The final feature of this phase of globalization that will be particularly relevant to analysis of the relationship between China's regions and globalization since the 1980s relates to the spatial structures of global political economy. This can be encapsulated in the metaphor of a 'network society', through which sociologist Manuel Castells analyses recent transformations in global political economy.[15] In this conceptualization, the dominant spatial configurations of political economy during recent decades have increasingly become those of global networks between metropolitan areas, with capital, information, technology, and people (elites more than workers) flowing across borders through the nodes in these networks – this conceptualization reflects the production networks described above too. An important feature of this network organization is an emphasis on the development of megalopolises or 'global cities', which will be touched later in this book. Importantly for our purposes, regions, too, are reformed during this phase of globalization. This 'shift in the allocation of resources from surfaces, or national spaces, to nodes in global networks'[16] is in contrast to an earlier period where bounded surfaces were the dominant spatial structure within which political economy was imagined and realized, reflected in the idea of the nation-state as the dominant political-economic 'power container'.

The network metaphor therefore highlights the extent of state transformation during this era, as well as the spatial limitations of globalization and the marginalization of territories and peoples not included in these networks. Of particular relevance to Chinese and other efforts to extend globalization, the proactive development of infrastructure to develop such networks can be seen as a form of 'spatial fix' to reduce spatial barriers to and speed up flows of capital, in particular through the building of transport networks to bring products to market.[17] Technology – or the technological subjugation of distance[18] – is an essential ingredient of this process through which 'space organizes time in the network society',[19] though technology should be seen as an enabling tool rather than driver of globalization.[20] This network society has also seen the rapid growth in financial globalization, which – as the post-2008 global crisis made clear – has increasingly driven developments in the global economy.

China and globalization

These features of globalization – the shift to a neoliberal 'hyperglobalization', the business revolution of production networks and global value chains, the dominant role of finance, and the spatial network structures these have created – are key to understanding how China's economic and social transformations since the 1980s relate to these processes of globalization. Many explanations of China's development during this period emphasize the role of policy change and developments within China, starting with the decision pushed by Chinese leader Deng Xiaoping in December 1978 to pursue 'reform and opening up', in response to the economic challenges facing China in the late 1970s. But the story of China's economic rise since then has been as much about the incorporation of China into a globalizing world economy as it has been about endogenous drivers in China.[21] As political scientist Edward Steinfeld puts it, China's 'political-economic transformation has been driven by globalization'; and crucially, China's 'linkage to the global economy ... took place when the worldwide economy itself was undergoing a profound transition', especially in the period from the early 1990s onwards.[22]

This idea is echoed in comments by professor of anthropology and geography David Harvey on Deng's move in 1978:

> We may never know for sure whether Deng was all along a secret 'capitalist roader' ... or whether the reforms were simply a desperate move to ensure China's economic security and bolster its prestige in the face of the rising tide of capitalist development in the rest of East and South-East Asia. The reforms just happened to coincide – and it is very hard to consider this as anything other than a conjunctural accident of world-historical significance – with the turn to neoliberal solutions in Britain and the United States.[23]

The key point here for thinking about the relationship between China and an emerging new phase of globalization – whether we call it neoliberalism, global capitalism, or hyperglobalization – was that this was a two-way relationship: in other words, under Deng Xiaoping, China was 'ready to enter the world, and the global economy was ready to integrate China'.[24]

In practical terms, the result was that the Chinese authorities set out to attract global capital in the form of foreign direct investment in manufacturing and assembly operations. They directed this into specific areas of the country concentrated initially along the coast, starting with four special economic zones (see Chapter 2), which led to the incorporation of parts of China's southern coast into regional and global production networks, which were themselves being extended by this incorporation. The attraction for global capital lay in the availability of low-cost labour and land in these parts of China, as well as the generally positive response from Chinese authorities in these areas. Over the following years this spurred a number of developments that would become strong features of the subsequent social and economic transformations in China, from the massive flows of migrants from inland provinces to factories along the coast, which brought tens of millions of Chinese workers into the global economy, to the environmental degradation that has resulted from much of this manufacturing activity.

This process of incorporation into the global economy was accelerated after Deng Xiaoping's 'southern tour' in early 1992 (see Chapter 2) reinvigorated the 'reform and opening up' process less than three years after the June 1989 tragedy in Tiananmen Square, and immediately after the collapse of the Soviet Union marked the end of the Cold War and with it the demise of socialist alternatives to capitalist development. This has been described by critics such as Chinese intellectual Wang Hui as an all-out embrace by the Chinese state of neoliberal policies and rejection of its socialist values. But other analysis has sought to account for the active role of the post-socialist state through hybrid concepts such as 'state neoliberalism', or compared the Chinese government's approach to the developmental state of other east Asian economies that experienced rapid growth in the latter part of the twentieth century.[25] The result was that China adopted 'a distinctive state-directed, yet marketized and globalized economic model'.[26]

A further key step in the intertwining of China's economy with globalization was mainland China's accession to the World Trade Organization in 2001, against the background of a judgment by Chinese policy makers that 'a growing array of goods were being produced in global rather than national production networks' and that there was therefore 'no viable alternative to becoming even more deeply involved in the globalizing economy'.[27] China's total trade then grew five times as fast as global trade to account for over 10 percent of global trade in goods ten years after its WTO accession, more than any other country; that proportion has since risen to over 14 percent. Still, the pattern of China's interactions with the global economy as substantially constituted by international and regional production networks continued: the Chinese Ministry of Commerce reported ten years after WTO accession that over half China's exports were from foreign-invested enterprises, and processing trade has consistently accounted for a substantial share of China's exports.

However, although companies in China have accounted for growing shares of world trade and manufacturing activity over the period from the 1990s, the value they have added in doing so, and their ability to control production, has remained highly limited. Rather, the pattern has been the import of components from elsewhere in the region and the export of assembled goods to markets further afield, often under the supervision of non-Chinese companies. This resulted in Chinese trade deficits with much of east Asia and surpluses with consumption-heavy economies in Europe and the US, though this has begun to change in recent years. The consciousness among Chinese policy makers that most of the country's economy is still at a low- to mid-point in global value chains explains much about current Chinese policy concerns to move up the value chain, and to develop enterprises that enable it to do this. Recent plans such as the 'Made in China 2025' industrial strategy, which seeks to develop China's innovation capacity and leadership in smart manufacturing, are attempts by the government to respond to these concerns.[28]

As noted above, globalization is not just about intensified economic interactions, and the political and ideological questions it raises were particularly important in Chinese considerations of how to approach globalization in the 1990s.[29] What emerged from these deliberations was a shift in elite discourse in China around 1997 from talk of China opening up to an 'international' world of nation-states, to a discourse of globalization and a globalizing world. This followed debates among policy elites around a number of specific questions about globalization: Did it actually exist as a phenomenon? What were its character and dimensions? What values were implicit in globalization? Was it advantageous to developing countries? Was it effectively controlled by the West and in particular the United States? Would it erode national cultures? In answer to these questions, Chinese Communist Party theorists constructed a positive picture of a globalizing world in which nation-states could retain their sovereignty and capacities despite the challenges to states brought by globalization.[30] In particular, they concluded that globalization was positive for China (and other developing countries), that it was value-neutral rather than 'capitalist', and that it was primarily economic in nature. Further, in this Chinese thinking, globalization – alongside a multipolar international order[31] – is an 'inevitable' trend, and many Chinese scholars continue to cite the two trends of multipolarity and economic globalization together.[32] The result was a view that managing globalization could be used to strengthen the Chinese nation, closer to seeing globalization as intensified interdependence between states rather than the creation of a 'world system' of new global political-economic structures that significantly challenge the nation-state.

Put another way, this can be seen as a degree of official Chinese resistance to the tendencies to fragmentation, decentralization, and internationalization of the state under globalization. These theoretical conclusions about globalization continue to inform official Chinese statements on globalization today. Almost all official references to globalization are to 'economic globalization'. China's approach is therefore rather more akin to the sort of Bretton Woods globalization that was a feature of the period up to the 1970s – as Dani Rodrik puts it, China and India 'chose to play the globalization game not by the new rules, but by Bretton Woods rules', though others have argued that in practice parts of the Chinese state often behaved in more neoliberal ways.[33]

Two related policy consequences of this political willingness to engage with globalization from the mid-1990s onwards were the idea that China should build its own transnational corporations and invest overseas, and begin to engage more proactively with global economic governance.[34] The development of meaningful levels of Chinese outward investment did not begin until around 2003. Initially this was predominantly driven by the need for resources, led by investments in extracting energy and minerals. In the 2010s the drivers and sectors have diversified, with access to technology, brands, and overseas markets accounting for a rising proportion of China's outward investment. Proactive engagement with global economic governance has also grown over the last decade, seen through China's approach to the G20 as it became the premier international grouping for discussions of the global economy after the global financial crisis struck. These two trends have come together more recently in the Chinese leadership's belt and road initiative and the launch of the Asian Infrastructure Investment Bank (AIIB).[35]

These earlier themes of China's approach to globalization feature in president Xi Jinping's much-hyped January 2017 speech at the World Economic Forum in Davos,[36] which was misleadingly reported in many outlets as a fulsome embrace of globalization. In line with the points outlined above, Xi referred throughout to 'economic globalization', which he described as a 'natural outcome of scientific and technological progress [which has] powered global growth and facilitated movement of goods and capital, advances in science, technology and civilization, and interactions among peoples.' He said that

> there was a time when China also had doubts about economic globalization, and was not sure whether it should join the WTO. But we came to the conclusion that integration into the global economy is a historical trend . . . It has proved to be a right strategic choice.[37]

However, Xi went on to qualify this with statements noting that economic globalization was a 'double-edged sword', which had strained both relations between capital and labour and the balance between efficiency and equity. He talked about the need 'to make the process of economic globalization more invigorated, more inclusive and more sustainable', and to 'act pro-actively and manage economic globalization as appropriate so as to release its positive impact and rebalance the process of economic globalization'. This need to 'adapt to and guide economic globalization' is precisely the sentiment that emerged from those theoretical Chinese debates about globalization in the 1990s. China's official engagement with globalization has always fallen short of a full embrace.

The picture of China's relationship with globalization painted so far in this chapter has focused very much on production processes, and the ways in which China's engagement with the global economy led it to become the 'factory of the world'. This engagement has not been reflected to the same extent in some other areas of globalization, in particular financial globalization.[38] The Chinese government has acted much more cautiously in opening up its financial sectors to foreign participants, and this remains the case more than 15 years after the PRC acceded to the WTO. In more general terms, the nature both of globalization and of China's relationship to it has continued to change, particularly in the aftermath of the post-2008 global financial crisis. These changes have included a growing role for Chinese outward investment, the emergence of companies such as the telecoms giant Huawei as Chinese corporations, which may have the potential to occupy global positions in their sectors similar to those played by companies from developed economies,[39] and declining growth in Chinese trade and dominance in global production networks as more elements of internationally disaggregated supply chains move elsewhere in the region and within China.[40] Chinese individuals have also featured prominently in the changing shape of globalized consumption, through their purchases of luxury goods produced by global brands, outbound tourism and shopping, and the number of Chinese students who have gone overseas to study.[41] In conclusion, as sociologist Ho-fung Hung put it in 2009:

> in 1978, the Chinese Communist Party undertook a bold step to rejuvenate China's economy by shifting it from a centrally planned system to a market system as well as by opening China to foreign investment. This dramatic transition coincided with the beginning of the profound transformation of global capitalism, a transformation commonly known as 'globalization,' which helped unleash decades of rapid economic expansion in China.
>
> After three whole decades of rapid growth, China is no longer an ordinary developing country or a formerly socialist economy struggling to *respond to* the challenge of global capitalism. It has become a geoeconomic and geopolitical heavyweight *reshaping* the structure of global capitalism.[42]

China's regions in an era of globalization

The rest of this book focuses not on the macro picture of China's dynamic relationship with processes of globalization, but on the role of China's regions. 'Regions' here refers to sub-national areas that generally encompass multiple administrative units (or parts thereof), and have some coherence and relevance from a political-economic or policy perspective or might be linked through geographical conditions or economic integration (not supra-national regions such as Asia or east Asia). These may be equivalent to provinces or municipalities, which form an important part of China's political system, but equally they may cross administrative boundaries or encompass multiple administrative units. This sub-national level has become more relevant with significant devolution since the

1980s, especially in economic and social policy, and national policy towards China's regions has developed in more complicated and sophisticated ways over recent years, especially since 2010. Although the national perspective is often dominant in studies of China, sub-national developments have therefore become increasingly necessary to understand the political economy of China, both domestically and externally.

From a theoretical perspective, this is a constructivist approach, which sees regions as 'socially constructed areas defined by state, supra-state, and societal agents, with shifting territorial, economic, and socio-political parameters'.[43] This approach allows space for a clear role for policy makers and administrative structures, but also for alternative factors to come to play in the way that regions are con-ceptualized. Indeed, there is potential for something of an imagined quality to regions; as Arjun Appaduri puts it, regions 'imagine their own worlds'.[44] And given the spatial structures of globalization, thinking about China's regional political economy should not always lead us to examine bounded surfaces, but also networks.

Although this book focuses on the development of China's regions in an era of intensified globalization since the 1980s, it is worth noting briefly that the external relations of China's sub-national regions have been highly relevant at various points in its history over the *longue durée*. China's history has been one of change in geographical extent and both internal and external orientation,[45] with both con-tinentalism and maritime orientation featuring at different times in history, legacies that have echoes today. During the westward-facing Tang dynasty (618–907 CE), 'silk roads' stretched across the Eurasian continent from Chang'an, today's Xi'an, a city that for much of the twentieth century has been considered in 'remote' western China. The Ming dynasty (1368–1644) grappled with tribal relations to its north, also part of the reason for the Qing dynasty's (1644–1911) subsequent expansion to the west under the Kangxi emperor.[46] By the middle of the nineteenth century, the focus had shifted eastwards to the coastal regions that were the primary targets of European incursions, and where the dynasty's external relations were managed differently from those with inner and central Asia. Late in the Qing dynasty, amidst debates over how to manage the empire's frontiers, the administrative system of provinces was expanded in the 1870s and 1880s to include previous frontier territories such as Xinjiang and Taiwan; and in response to external developments, the Qing's localized and devolved frontier policy, which had often been effected through local power holders in frontier territories, was gradually replaced by a more centralized, comprehensive policy that 'conceived of a single hierarchy of imperial interests framed in reference to a unified outside world'.[47]

This process was part of a wrenching transformation from Qing empire to Chinese nation-state, which culminated in the establishment of the Republic of China in 1912, followed by a period of limited success in integrating national territory, and the subsequent Japanese invasion and occupation from the 1930s to the end of World War II. To the extent that regional policy existed as a coherent concept during the Republic era it was primarily about integrating and controlling as much as possible of the territory formally claimed by the Nationalist government. From 1938 it was also about using the security provided by more remote locations in southwest China to allow Chiang Kai-shek's government to survive, at the same time as the Communist Party was using remote regions of the country to build its own strength. During these years, trends in regional political economy were diverse, from pre-war cosmopolitan Shanghai to places such as Yunnan in the southwest, which retained some linkages with neighbouring societies outside China's borders.[48]

In the first decades of the People's Republic the emphasis was on state building and self-sufficiency and there were minimal external linkages of China's regions. This all began to change after 1978 when 'reform and opening up' led to the gradual integration of coastal China with the global economy. By the second decade of the twenty-first century, there were diverse global interactions across different parts of China, from the globally competitive information technology enterprises found in the southern city of Shenzhen to the strategy of Ningxia Hui Autonomous Region in the northwest to develop economic and commercial linkages with Muslim societies outside China. To put these developments in context requires a detailed account of the development of China's regions and regional policy.

Notes

1 Arif Dirlik. *Global modernity: modernity in the age of global capitalism*. Boulder, CO: Paradigm, 2007, p. 1.
2 Nick Knight. *Imagining globalisation in China: debates on ideology, politics and culture* Cheltenham, UK: Edward Elgar, 2008, pp. 51–53.
3 Angus Cameron and Ronen Palan. *The imagined economies of globalization*. Thousand Oaks, CA: Sage, 2004, p. 2. Their book sets out to critique this understanding.
4 Nicola Phillips. 'Power and inequality in the global political economy', *International Affairs*, 93: 2, 2017, pp. 429–444.
5 For example: Jeffrey Frieden. *Global capitalism: its fall and rise in the twentieth century*. New York: W. W. Norton, 2006.
6 Dani Rodrik. *The globalization paradox: democracy and the future of the world economy*. New York: W. W. Norton, 2011, p. xvii.
7 In the literature this has been characterized in numerous ways, whether in terms of a turn to neoliberalism, late capitalism, or a post-Fordist regime of capital accumulation. For examples, see: David Harvey. *A brief history of neoliberalism*. Oxford, UK: Oxford University Press, 2005 and Ho-fung Hung, *China and the transformation of global capitalism*. Baltimore, MD: The John Hopkins University Press, 2009.
8 Milivoje Panić. *Globalization: a threat to international cooperation and peace?* Basingstoke, UK and New York: Palgrave Macmillan, 2011, p. 16.
9 For a more detailed exposition of these processes see Chapter 1 of Panić, *Globalization* (see note 8). This section also draws on the descriptions of globalization in: Edward Steinfeld. *Playing our game: why China's rise doesn't threaten the West*. Oxford, UK: Oxford University Press, 2010 and Peter Nolan. *Is China buying the world?* Cambridge, UK: Polity Press, 2012.
10 Cited in Christopher M. Dent. *East Asian regionalism*. London and New York: Routledge, 2008, p. 46.
11 Neil Brenner, Bob Jessop, Martin Jones, and Gordon MacLeod, eds. *State/Space: a reader*. Malden, MA and Oxford, UK: Blackwell, 2003; Shahar Hameiri and Lee Jones

'Rising powers and state transformation: the case of China.' *European Journal of International Relations*, 21: 1, 2016, pp. 72–98. For 'paradiplomacy', see: Alexander Kuznetsov. *Theory and practice of paradiplomacy: subnational governments in international affairs*. London and New York: Routledge, 2015.
12 Ho-fung Hung, ed. *China and the transformation of global capitalism* (see note 7).
13 Nolan, *Is China buying the world?* (see note 9). For discussion of Apple, see: Jenny Chan, Ngai Pun and Mark Selden. 'The politics of global production: Apple, Foxconn and China's new working class', *New Technology, Work and Employment*, 28: 2, 2013, pp. 100–115.
14 Caroline Freund and Dario Sidhu. 'Global Competition and the Rise of China', Peterson Institute for International Economics Working Paper, Feb. 2017, https://piie.com/publications/working-papers/global-competition-and-rise-china.
15 Manuel Castells, *The rise of the network society*. Oxford, UK and Malden, MA: Blackwell, 2010.
16 Dirlik, *Global modernity* (see note 1), p. 24.
17 David Harvey. *Spaces of capital: towards a critical geography*. Edinburgh, UK: Edinburgh University Press, 2001, p. 123.
18 John Garver. 'Development of China's Overland Transportation Links with Central, South-west and South Asia.' *The China Quarterly*, 185, 2006, p. 18.
19 Castells, *The rise of network society* (see note 15), p. 407.
20 Knight, *Imagining globalisation* (see note 2), p. 17.
21 David Zweig. *Internationalizing China: domestic interests and global linkages*. New York: Cornell University Press, 2002.
22 Steinfeld, *Playing our game* (see note 9), pp. 16–17.
23 Harvey, *A brief history of neoliberalism* (see note 7), p. 120.
24 Andrew J. Nathan and Andrew Scobell. *China's search for security*. New York: Columbia University Press, 2012, p. 245.
25 Wang Hui. *China's new order: society, politics, and economy in transition*, edited by Theodore Huters. Cambridge, MA: Harvard University Press, 2003. For 'state neoliberalism', see: Alvin So and Yin-wah Chu. *The global rise of China*. Cambridge, UK: Polity, 2016.
26 Nathan and Scobell, *China's search for security* (see note 24), p. 257.
27 Nicholas Lardy and Lee Branstetter. 'China's embrace of globalization', NBER Working Paper 12373, July 2006, www.nber.org/papers/w12373.pdf, p. 21. Taiwan also joined as a separate customs territory at the same time; Hong Kong was already a member in its own right.
28 Mercator Institute for China Studies. 'Made in China 2025: the making of a high-tech superpower and consequences for industrial countries', December 2016, www.merics.org/en/merics-analysis/papers-on-china/made-in-china-2025/.
29 This paragraph is based on excellent in-depth treatment of these questions in Knight, *Imagining globalisation* (see note 2).
30 Subsequent references to the 'Party' in this book are to the Communist Party of China.
31 A multipolar international order is one with many 'poles', or power centres, in contrast to a unipolar order.
32 For example: Cui Liru. 'Toward a multipolar pattern: challenges in a transitional stage'. *China–US Focus*, April 2014, www.chinausfocus.com/foreign-policy/toward-a-multipolar-pattern-challenges-in-a-transitional-stage/.
33 The quotation is from Rodrik, *The globalization paradox* (see note 6), p. xviii. For criticism see Hui, *China's new order* (see note 25), also: Ho-fung Hung. *The China boom: why China will not rule the world*. New York: Columbia University Press, 2016.
34 Knight, *Imagining globalisation* (see note 2), Chapter 7.
35 See: Tim Summers. 'Thinking inside the box: China and global/regional governance'. *Rising Powers Quarterly* 1: 1, pp. 23–31.
36 Xi Jinping. 'Jointly shoulder responsibility of our times, promote global growth', speech delivered at World Economic Forum Annual Meeting, Davos, 17 Jan. 2017, http://news.xinhuanet.com/english/2017-01/18/c_135991184.htm.

37 For a good summary of the debates in China about the benefits of WTO accession between 'pro-globalizers' and their opponents, see: Mark Beeson and Fujian Li, *China's regional relations*. Boulder, CO and London: Lynne Rienner, 2014, pp. 20–21.
38 Paola Subacchi. *The people's money: how China is building a global currency*. New York: Columbia University Press, 2016.
39 Peter Nolan, *Chinese firms, global firms: industrial policy in the age of globalization*. New York: Routledge, 2014, pp. 152–157.
40 Cristina Constantinescu, Aaditya Mattoo, and Michele Ruta. 'The global trade slowdown: cyclical or structural?' IMF Working Paper, 2015, www.imf.org/external/pubs/ft/wp/2015/wp1506.pdf.
41 LeAnne Yu. *Consumption in China: how China's new consumer ideology is shaping the nation*. Cambridge, UK: Polity, 2014.
42 Hung, *China and the transformation of global capitalism* (see note 7), p. vii (emphasis in original).
43 Mark Selden. 'Economic nationalism and regionalism in contemporary East Asia'. *The Asia–Pacific Journal*, 10: 43, October 2012, http://japanfocus.org/-Mark-Selden/3848.
44 Cited in: Carolyn Cartier. 'Origins and evolution of a geographical idea: the macroregion in China'. *Modern China*, 28: 1, January 2002, p. 124.
45 Valerie Hanson. *The open empire: a history of China to 1600*. New York and London: W.W. Norton, 2000.
46 Peter Perdue. *China marches West: the Qing conquest of Central Eurasia*. Cambridge, MA: The Belknap Press of Harvard University Press, 2005.
47 Matthew Mosca. *From frontier policy to foreign policy: the question of India and the transformation of geopolitics in Qing China*. Stanford, CA: Stanford University Press, 2013, p. 203. See also: C. Patterson Giersch. '"Grieving for Tibet": conceiving the modern state in late-Qing inner Asia'. *China Perspectives*, 3, 2008, pp. 4–19.
48 Tim Summers. *Yunnan: a Chinese bridgehead to Asia: a case study of China's political and economic relations with its neighbours*. Oxford, UK: Chandos, 2013, pp. 40–46.

2
REGIONAL POLICY IN CONTEMPORARY CHINA

A historical overview

Understanding the development of sub-national regions in China and their role in China's engagement with globalization from the 1980s first requires an introduction to the formal political and administrative organization of China's territory. One feature of the country's political system is a hierarchical structure of local government from the national level through province (or provincial-level municipality and autonomous regions) to prefecture/city, county/city, township and village (see Figure 2.1).[1] Official institutions of the Party-state in China can be found with some variation at each level, from Party committees to government departments and people's congresses (villages are formally self-governing, though they retain Party committees). This hierarchy of sub-national levels of government plays an important role in the development and implementation of regional policy in China, as well as in broader management of the economic and society. It is infused by the Communist Party, at the formal apex of which sits the Central Committee (205 members and 171 alternates from 2012), with key strategic decisions being made by its Politburo of 25 or Politburo Standing Committee of seven members. The Party's General Secretary since November 2012, Xi Jinping, is also state president and chair of the Central Military Commission, and following the 19th National Congress of the Communist Party in October 2017 he looks set to retain these positions at least until 2022.

The geography of these political structures has shaped economic activity, and led to the phenomenon of 'administrative region economies' in China, whereby administrative boundaries and the competition between administrative units for resources acts as a constraint on cooperation and coordination across administrative units.[2] At the same time, analysis of the development of China's regions cannot just be based on these administrative structures, but needs to take account of numerous organic social and economic trends that do not necessarily map onto them.[3] Language or culture often cut across provincial or other local boundaries

– as cultural geographer Tim Oakes suggests, 'China has a tradition of place-based identities that have seldom, if ever, corresponded with provincial-administrative boundaries'.[4] In some cases, administrative boundaries may even have been designed to reduce the potential political or economic strength of cultural or economic regions. And scholars have for some time been arguing that 'the current economic regions are by no means the same as the traditional provinces, not only in geographical terms, but also administratively, politically and economically'.[5] In sum, although provinces are the 'most likely regional sub-divisions to hold local power', they 'might no longer have the size and scope to provide effective economic and regulatory coordination, thus creating a role for greater regions'.[6]

The development of regional policy, especially in recent years, shows some efforts by policy makers in China to devise and develop regional economies that go beyond provincial administrative structures.[7] These range from the major areas of central and western China that were brought together as the focus of the Third Front policy in the 1960s, to the definition of a new concept of 'western China' under the Develop the West programme launched at the turn of the century, and more recent policy goals such as building a 'Yangzi river economic belt' (see below). Here, the concept of 'macro-region' is useful in describing regional constructs that cross provincial and other administrative boundaries, and that can therefore 'serve as important frameworks of regional analysis'.[8] It also echoes anthropologist G. William Skinner's landmark demarcation of late Qing dynasty China into nine non-contiguous macro-regions based on watersheds and marketing systems, somewhat akin to 'natural economic territories'.[9]

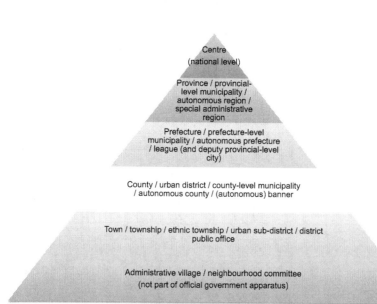

FIGURE 2.1 China's local government system

In the period since the 1980s the role and influence of sub-national governments in China's economic and social development has grown. When it comes to policy making and implementation, an important question is therefore the changing relationships between the different levels of government. Many studies of centre-local relations in China look at policy implementation in terms of principal-agent relations, or evaluate the extent of autonomy that provincial and other subnational administrative units have within the context of policies set out by the central authorities.[10] Although the PRC is a unitary state, the ability of the centre to implement its policies remains limited, though where these limits lie is a matter of debate and varies across place and issue. The central Party-state bureaucracy is relatively light, and some 70 percent of Chinese bureaucrats are found at the sub-provincial levels.[11] Further, although policy statements are promulgated and information and data collected and disseminated through the geographical hierarchy from central to local levels of government, policy implementation is made more complicated by a corresponding vertical system of functional (sectoral) departments at national, provincial, and lower levels in the hierarchy. The Chinese political-administrative structure is therefore best envisaged as a vertical-horizontal matrix through which state power is distributed in complex ways; this informs policy making and implementation in most areas, for example in the regulation and management of business.[12]

The centre-local framework tempts us to look at research into China's regions or provinces as a series of case studies that will shed light on generalizable dynamics of the relations between national-level authorities and their local counterparts. But as Harvard scholar Meg Rithmire argues in a discussion of the implications of 'new regionalism' in China, understanding locally-determined change (and therefore sub-national dynamics) requires a rejection of homogeneity in favour of the examination of local variables, path dependency and the diverse political-economic and institutional history of various regions, their resource endowments and government decisions, leadership and capacity – all without forgetting the 'lurking power of the centre'.[13] In other words, every part of China is different, and there are no simple patterns that can be determined across the country. The picture that emerges is therefore one of 'structured uncertainty' and ambiguity,[14] based on this substantial diversity and variation across the country. Indeed, if there is one thing that above all else characterizes China's 'regions', it is diversity. The differences not just in economy, but in geography, climate, food, language, customs and more across China's territory offer a vast canvas for research; the diversity in the global interactions of China's regions is another part of this picture. One conclusion from the emphasis on regional diversity is the possibility that specialization and diversification, and the unique set of capabilities of each region, can enable parts of China to play different roles at various stages within a fragmented global economy and thereby enhance the overall sustainability of China's economic development.[15]

With these points in mind, this book attempts to move away from the idea of regions as case studies of wider theorizable phenomena and looks at the regions it considers in their own right and in the context of their own historical, national, and global development to draw conclusions about the roles they have in the past and might in the future play in China's relationship to globalization. There is evidence too that the Chinese government itself approaches regions in this way; for example, the National Development and Reform Commission's annual report of March 2016 talks about supporting 'Ningxia, Guizhou, and other parts of the western region in developing open inland economies, *each with its own distinctive features*', echoing a longer-standing theme of policy making in China.[16]

This book also builds on studies that look at sub-national and regional political economy and regional policy, including important collections of research papers on themes related to China's provinces in reform, and books on individual provinces.[17] This political economy literature sits alongside studies in anthropology or culture, many of which delve below the provincial scale to look at smaller regions or peoples, often among the ethnic minority populations of western China.[18] There are also good historical accounts of a number of provinces, such as Sichuan.[19] There is plenty written on Tibet and Xinjiang, some of which directly brings in the impact of global interactions, though often in terms of politics as well as political economy.[20]

The role of subnational governments and the devolution of greater power to provinces has not just related to domestic policy implementation, but has had implications for China's global interactions. This includes the devolution to provincial governments of authority in managing foreign trade and some other aspects of foreign relations, such as dealing with foreign governments and even border demarcation with Vietnam.[21] There is an ongoing debate over the extent of provincial or local agency in these interactions, in other words how much space provincial governments or other institutions have to set the agenda, to implement policies in ways that reflect local conditions or interests, and to form external economic and commercial ties.

Research conducted to date has produced a range of conclusions. Some scholars have highlighted the growing international activity of provincial and municipal governments, but concluded that their participation in China's foreign affairs is still 'limited' and 'indirect'.[22] Others argue that, although some devolution in foreign trade and other areas of foreign relations has taken place, the outcomes are not in conflict with national goals.[23] In my earlier book on the southwestern province of Yunnan, I emphasized more the scope for provincial agency in global interactions, but within the frameworks of national policy.[24] Other research has suggested that local players are among many new foreign policy actors who have been injecting a wider range of often conflicting interests into the making of Chinese foreign policy.[25] These arguments are pushed further by those who see the 'novel and epochal' transformation of the Chinese state under the influence of globalization meaning that it no longer makes sense (if it ever did) to look at the Chinese state as a unitary actor in external relations.[26] Other scholarship has taken as its focus regions that are not only not primarily defined by the provincial scale, but also cross national borders to create transborder subregions.[27]

Self-sufficiency and redistribution: from the 1950s to the Third Front

On the establishment of the People's Republic of China in October 1949 the main focus of regional policy was domestic. One of the first challenges facing the Communist Party was to gain effective control over its territory. In parts of western China this was not achieved until after October 1949, and in some border and ethnic minority areas, such as Xishuangbanna in Yunnan province, the process of state penetration extended well into the 1950s. Regional policy reflected a dual desire to integrate regions into the new nation and to enhance integration within regions themselves. Six Greater Administrative Regions (GARs) were established, up to 1954, to facilitate political and military control, and oversee economic cooperation and integration between and within regions.[28] Investment in industry was the main tool of regional policy during this period.

Another state-building priority related to so-called 'ethnic minority' populations. As well as the classification of ethnic groups, this involved the demarcation of parts of the country as 'autonomous regions'. At the provincial level, the first had been established in 1947 in Inner Mongolia. The next was the Xinjiang Uyghur Autonomous Region in 1955, followed by the Guangxi Zhuang and Ningxia Hui Autonomous Regions in 1958, and much later the Tibet Autonomous Region in 1965. Below the provincial level, numerous autonomous prefectures and counties were established, especially to the west: 27 out of 30 autonomous prefectures and 83 of the 120 autonomous counties are in western China. Preferential policies for ethnic minorities followed, some linked to an individual's ethnic status, such as less strict birth control policies and quotas for university admission, while the governments of autonomous areas retain more taxes for local spending and enjoy subsidies for development.[29]

Alongside these questions of regional demarcation and administrative structures, the PRC era began with 'substantial gaps in development levels among the provinces and regions',[30] with Shanghai and to a lesser extent Tianjin, Beijing, and Liaoning, well ahead in terms of GDP per capita. This drove one of the major issues in China's regional policy for the coming decades, trying to achieve a better economic balance across the country, and in particular between coastal and inland regions.[31] Policy makers' initial concerns about the 'irrational' concentration of some 70 percent of industry in coastal provinces were addressed in the first five-year plan (1953–1957) by placing 472 of 694 planned large industrial projects in China's interior.[32] Mao Zedong later commented in 1956 that 'we have not laid enough stress on industry in the coastal regions', and that although over 90 percent of new factories should be set up in the interior,[33] the strengths of coastal industry should be used to support industrial development in inland China. The proposed second five-year plan, for 1958–1962, reflected more of a coastal emphasis than the first, but was quickly overtaken by the damaging 'great leap forward' from 1958 to 1960, and a much greater emphasis on self-reliance, not just at the national level, but locally and regionally too. In 1958 a macro-regional administrative structure was revived, this time consisting of seven regions focusing on economic coordination and 'the aim of developing economic linkages among different provinces',[34] without the political and military dimensions of the earlier GARs.

Although these policies were formulated around a domestic economic desire to distribute industry in a way that would provide more equitable regional development, the external context was also relevant. While the new PRC was engaged in conflict on the Korean peninsula from 1950–1953, inland China was seen as more secure, as it had been for Chiang Kai-shek's Nationalists from the late 1930s to 1945, though after the Korean War, 'the coastal regions were no longer considered strategically vulnerable'.[35] But the period from 1956 was marked by China's growing isolation, as relations with Moscow deteriorated, culminating in the Sino–Soviet split. This was at least partly behind the policy emphasis on self-reliance and decisions on the location of heavy industry.

The Soviet Union and the United States were important drivers for a subsequent major policy with significant regional implications, the so-called 'Third Front' movement, which began in 1964. Put forward by Lin Biao, who was at that point emerging as heir apparent to Mao, the plan was a response to perceived strategic threats from the United States, which intensified its military action in Vietnam, and the USSR, which later in the decade stationed troops in Mongolia and enunciated the Brezhnev doctrine on its claimed right to use military force to maintain its communist rule in other socialist countries. The fears of the Chinese leadership should not be assumed to be irrational, following the Cuban missile crisis and at a time when the PRC continued to work on the development of nuclear capability without Soviet assistance, against the wishes of the US and USSR who 'hinted broadly that they might launch limited nuclear strikes to destroy China's embryonic nuclear capacity';[36] 1964 was the year when China's successful test took place.

Stimulated by these strategic concerns, the essence of the Third Front was to shift heavy and military industries inland and invest in railway infrastructure, rather than expose China's industry by relying on the coast. Analysis by economist Barry Naughton splits this into three areas or phases: the southwest, dominant during 1965–1969 when the main response was to perceived threats from the US; central China, mainly from 1969 onwards when the strategic worries turned more to the USSR; and the northwest, where the state continued military-industrial development that had already begun before 1964. This categorization highlights that the Third Front cut across both provincial boundaries and the timelines of five-year plans, which makes evaluation of its impact more difficult.[37] Rather than being based on provincial structures, the locations of Third Front investment were decided on the basis of geography or topography, concentrated in the rugged and mountainous terrain of parts of central and western China, with points at the boundaries of provinces, such as Panzhihua in southern Sichuan, playing key roles. The administrative structures of the Third Front reflected this, with direct 'command points' being set up to deliver projects outside the existing ministerial or provincial governance structures. Central planning caught up in the third five-year plan (1966–1970), which reflected this 'strategic reallocation of China's industrial investments' towards inland China.[38] The intensive phase of the Third Front came to an end in 1971, the point of rapprochement with the US symbolised by president Nixon's visit to China in February 1972, and the not-unrelated demise of Lin Biao,[39] though ongoing nervousness over relations with the Soviet Union meant that some Third Front projects continued beyond then.[40]

The impact of the Third Front on China's economy and its regions was substantial. Naughton is highly critical of the plan for its economic inefficiencies, describing its 'negative impact on China's economic development [as] more far-reaching than the disruption of the Cultural Revolution' and suggesting that investment would have been better targeted elsewhere.[41] A more recent revisionist interpretation emphasises the point that the Third Front was not intended to maximize economic growth but to secure national defence, and that the construction of railways, which was a central part of the strategy, left a number of important legacies for the later development of regions involved in the Third Front. In particular, it expanded the rail network westward by adding ten inter-provincial lines and integrated much of inland China into the national system. Indeed, the completion in 1965, 1966, and 1970 of Chongqing–Guiyang, Kunming–Guiyang, and Chengdu–Kunming lines respectively serves as a reminder of the infrastructure legacies of the programme, and of what was achieved through state planning even during the otherwise tumultuous early years of the Cultural Revolution.[42]

The Third Front demonstrates the impact of the global context on regional policy and regional development in China. From the early 1970s, and the post-Third Front fourth and fifth five-year plans (1971–1975 and 1976–1980), however, the focus of policy shifted back to coastal provinces. These years also saw the beginnings of economic opening. China's first large-scale technology import from the west, sheet rolling mills and chemical fertilizer plants approved in January 1973, may even have been a result of the Third Front.[43]

However, the redistributive policy emphasis on inland China from the 1950s to the early 1970s did not result in a narrowing of economic disparities between provinces, at least when measured in terms of provincial GDP per capita,[44] suggesting that it was not the case that when it comes to regional policy a 'notion of "equal poverty" [was] prevalent during the Mao-dominated era'.[45] There were changes in provincial ranking measured by GDP per capita, most notably with the post-1949 capital Beijing moving ahead of Tianjin and the northeast by 1978.[46] In the early 1980s, though, the most developed regions were still along the coast, especially Shanghai, Beijing, Tianjin, Jiangsu, and the north-eastern provinces of Liaoning and Heilongjiang.[47]

'Reform and opening up' and uneven regional development

A new phase of regional policy therefore began from the 1970s onwards, featuring the development of coastal areas and engagement with the global economy, and characterized by 'uneven development'. Although policy tilted back towards the coast in the early 1970s, the decisive point came in December 1978 with 'reform and opening up' as the Party put economic development at the heart of its approach to governing China and began the process of China's incorporation into capitalist globalization. This involved an 'explicit regional policy, in which some regions [were] positively encouraged to become wealthy before others'.[48] As Deng Xiaoping put it in March 1980, China should use its 'comparative advantages, avoid using our disadvantages and accept the fact of economic disparities . . . some people *and some regions* should be allowed to get rich first and in the end everyone will get rich'.[49] As well as embracing regional inequality in the interests of aggregate national development, Deng's reference to 'comparative advantage' points to a shift in the main idea underlying policy from self-sufficiency towards something that more closely resembles classical economic principles. Indeed, criticisms of pre-1980s regional policy from policy advisers had included the argument that it had paid insufficient attention to coastal industry, and that the idea of self-sufficiency weakened production and economic diversity.[50]

The regions that were envisaged as the focus of opening up and more rapid wealth accumulation were along China's coast. This did not necessarily reflect either existing levels of economic and commercial development or factor and resource endowments (in spite of the reference to 'comparative advantage'), but a combination of experimental policy measures and potential for integration into the global economy and 'taking advantage of the specialisation opportunities of international trade'.[51] These two elements came together in the establishment in 1980 of four special economic zones (SEZs) along the south coast, in Shenzhen, Zhuhai and Shantou (Guangdong province) and Xiamen (Fujian province), in a move apparently deliberately drawing on the earlier experience of South Korea and Taiwan in integrating into regional production networks. Shenzhen's proximity to Hong Kong, and through it to the wider global economy, gave it a particular advantage in taking forward China's opening up. The SEZs 'provided investors with preferential policies in terms of tax holidays and exemptions, access to land and infrastructure, and special privileges in import and export',[52] and in the 1980s both Guangdong and Fujian were given preferential fiscal policies whereby they were allowed to retain more of their local revenue when a new general revenue sharing system was devised between the central government and provinces.[53]

The SEZs were followed in May 1984 by the opening up of 14 more coastal cities (with fewer preferential policies than the four SEZs) and of Hainan island, which would become a province in 1988 – in effect a 'provincial SEZ',[54] as well as the opening of development areas around the Pearl River Delta (Guangdong), Yangzi River Delta (Shanghai and neighbouring areas) and Min River Delta

(Fujian). The coastal emphasis was reflected in the sixth five-year plan (1981–1985) and the coast's vanguard role in China's opening up can be seen in statistics for foreign direct investment, 92 percent of which went to coastal provinces in the period from 1979–1987.[55] The priority given to coastal China was enhanced and made more explicit after the key thirteenth national Party congress in October 1987, when Zhao Ziyang pushed forward with a 'coastal development strategy',[56] designed to accelerate the development of an export-oriented economy by opening the seaboard from the northeast down to Guangdong. This would be through developing labour-intensive processing industries that would 'import raw and semi-finished materials from the international market and then export the finished products', or place 'both ends outside (*liangtou zaiwai*)' of China.[57] This was effectively an accelerated incorporation of coastal China into a global capitalist economy, though the Party squared this theoretically by deploying an official descriptor of China as being in the 'primary stage of socialism'.[58] The implementation of this strategy did not prove straightforward, and constraints included the challenges in dealing with the ongoing imbalances between coastal and inland areas.[59]

This focus on developing coastal China through integration into the global economy was not completely at the expense of inland areas. The sixth five-year plan (1981–1985), which still differentiated policy between coastal and inland China (and included particular measures for ethnic minority regions), was followed by a more nuanced demarcation in the seventh plan (1986–1990) into coastal, central, and western China.[60] This plan enshrined the idea of comparative advantage in regional policy,[61] though policy makers expected that the coastal focus would increase regional economic gaps, but sought to compensate by talk of a 'ladder-step doctrine' through which growth would eventually permeate less developed regions from west to east (echoes of the neoliberal 'trickle down' ideology elsewhere in the 1980s), and by looking for opportunities to support inland provinces. This was the motivation for a tour of southwest China in 1984 by then Party General Secretary Hu Yaobang, during which he encouraged these provinces to engage in opening up. In 1984, the central authorities had also encouraged 'economic networking among jurisdictions',[62] leading to a proliferation of regional structures that crossed administrative boundaries, including in the southwest where a group of provinces began to meet regularly to discuss ways of enhancing development.[63]

When Zhao Ziyang was pushed from power in the wake of the bloody suppression of the social and democratic movements of 1989, the coastal development strategy was effectively put on hold. But in 1992, Deng Xiaoping's 'southern tour' of Shenzhen, Wuhan, Shanghai, and Zhuhai restarted reforms and further accelerated China's coastal development and the process of China's incorporation into the global capitalist economy.[64] As in the 1980s, China's incorporation into globalization was actually the incorporation of *parts of coastal China* into the global economy, creating 'a series of port city-based boom towns tied to the world economy'.[65] In the 1990s, though, the geographical focus of this shifted further along the coast, with a particular emphasis on developing Shanghai (the Pudong Open Area had been established in Shanghai in 1990). Whereas the Pearl River Delta had dominated China's opening up in the 1980s, the 1990s saw the economic emergence and global engagement of the Yangzi River Delta, followed by the Bohai Bay area around Beijing and Tianjin to the north.

The development of these parts of coastal China and their global production networks would not have been possible without some connections to inland China. In particular, the growth of migration within China, primarily from inland provinces such as Sichuan towards the coast (there was also migration to Xinjiang in the northwest from the 1990s), facilitated the introduction of tens of millions of Chinese workers into the global economy. In the first half of the 1990s migrant flows towards Guangdong clearly dominated, but from the later part of the decade and into the 2000s flows became more complex and diverse along the coast.[66] This low-cost, literate, and hard-working labour force became a big part of China's 'comparative advantage' in the global division of labour, though they often did not enjoy proportionate access to the benefits that accrued to national aggregate economic growth.

By the 1990s, therefore, parts of coastal China had become integrated into global and regional production networks as China's economic development was spurred by the development of domestic regions with growing global economic linkages. The implications of the more peripheral areas of the country being the most globally integrated was debated inside and outside China, with noted China scholar Gerald Segal suggesting that the combination of the growing 'pull of . . . outside forces' on China's regions along with limited central government control (especially over provinces such as Guangdong) under conditions of economic devolution and fiscal weakness posed risks to China's integrity, even 'possible, albeit unlikely, breakup'.[67]

By the mid-1990s the extent of domestic regional economic disparities resulting from the emphasis on coastal development was becoming more apparent. Influential in this regard was research by academics Hu Angang and Wang Shaoguang,[68] which argued that regional disparities could pose a threat to political stability and national unity if they grew too large. These concerns had already begun to surface in response to the coastal development strategy,[69] incidentally questioning the idea that there was a coastal reform-oriented and inland conservative coalition.[70] Since the 1980s there had been lobbying by officials and researchers from inland provinces, such as Yunnan,[71] to pressure the centre to devise policies that would allow them to develop in the ways that coastal China had done. Part of the centre's challenge was fiscal: the central government's revenue had declined (in 1995 government revenue hit a low point of 10.7 percent of GDP[72]), and provincial governments such as that in Guangdong benefited disproportionately from their growth in the 1980s. Fiscal reforms pushed through in 1994 by then vice premier Zhu Rongji began to redress this situation, with a new formula for sharing revenue, which would strengthen the centre's ability to redistribute between regions. Possibly prompted by remarks Deng had made first in 1988 (when the coastal development strategy was being pushed forward) and again in 1992 about the need for a second stage of regional development for inland China,[73] material in the ninth five-year plan (1996–2000) stressed the importance of more balanced development of regional economies and narrowing gaps, as well as marking out seven trans-provincial regions for coordinating development and combating the tendency to administrative regional economies by aiming to 'break down the tendency for the three regions [east, central, west] to become relatively unconnected independent economies'.[74] It also promoted industrial investment, infrastructure development, and loans by international financial institutions in central and western China.[75] This all led up to an important shift in regional policy at the turn of the century, driven less by engagement with the global economy than concerns over domestic territorial integrity.

'Develop the West' and coordinated regional development

Around the turn of the century, PRC regional policy shifted from its phase of uneven development to one of 'coordinated development' between various regions.[76] This was marked by a major policy framework to 'open up' or 'develop' western China (henceforth 'Develop the West'), announced in 1999. The main aims of Develop the West were to speed up development in western China, which was defined from the policy's formal launch in 2000 as twelve provincial-level administrative units, with Guangxi and Inner Mongolia added to the nine (later ten) province-level units that had earlier been identified as western China in the ninth five-year plan. The programme was broad and exhortative in scope, and although there were new institutional structures at the national level to oversee it, provincial and ministerial lines remained the most influential when it came to policy implementation. Specific measures included tax breaks, fiscal transfers, increased lending, and an emphasis on infrastructure investment.[77]

There were also political and social elements to the programme, responding to the earlier concerns that national integrity could be undermined by development gaps; in particular the programme encompassed the Party-state's ethnic minority work, which partly explains the inclusion of the Guangxi and Inner Mongolia autonomous regions in the newly-demarcated western China; one consequence of the rhetoric around 'western development' was to create the impression that western China was 'backward' and constituted by ethnic minorities.[78] While the aim was to enhance development levels, the programme also built on a trend established earlier in the 1990s of exploiting resources in western China to feed economic growth in the coastal regions, for example through long-standing projects to 'send western electricity to the east'.

A further consequence was to highlight other geographical areas in China where development was not as fast as it had been on the coast. This led to subsequent development programmes at the macro-regional level covering the northeast and then the central provinces, though the next regional programme – 'revive the

26 A historical overview

northeast and other old industrial base areas' from 2003 – had a broader target in addressing legacies of state-owned heavy industry and the fear of social unrest, problems that were starkest in the northeast but not limited to that part of China.[79] The third macro-regional policy programme covered six provinces of China's 'central regions', responding in particular to concerns about weakness relative to the coast in terms of human capital, infrastructure and transport, low levels of economic efficiency, policy disadvantages, and collective protests. As premier Wen Jiabao put it in March 2004, 'accelerating the development of central China is an important aspect of our endeavours to ensure well-balanced development of regional economies'.[80]

One result of this series of policy frameworks was the effective demarcation of China into four macro-regions, what one China scholar has called 'four officially-recognised regions',[81] and this structure has continued to inform policy making. By the end of the 2000s, there was a sense that there had been some closing of inter-regional social and economic gaps, with the National Development and Reform Commission reporting annual average GDP growth in western China slightly faster than the national average, an increase in rail and road networks in western China of 1.5 and 2.8 times respectively, rises in urban and rural incomes of 2.7 and 2.3 times respectively, and a reduction in numbers in 'absolute poverty' from 57 million to 23.7 million. But the government also noted that western China's GDP per capita was 45 percent and disposable income and rural income 68 and 53 percent respectively of that in coastal regions, and transport density was 'half the national average'. Policy makers therefore concluded that more emphasis on western development was needed, resulting in a July 2010 high-level Party and government meeting chaired by Hu Jintao to launch a 'new round' of develop the west.[82] While the first phase had mainly focused domestically, an important new element of the 'new round' was the aim of developing western China into a 'bridgehead' from China to the rest of Asia, to take forward 'opening to the west', a concept to which we will return below.

New themes in regional policy from 2010

Up to this point, much of regional policy was driven by both ensuring continued economic growth and development, often led by particular regional growth poles, and finding the optimal balance in economic and social development between regions,[83] whether in the Mao-era swings between coastal and inland policy priorities, or the ideas behind Develop the West. However, a newer strand of regional policy came to the fore in 2010 with the publication by the central government of a lengthy policy document on 'major functional zones'.[84] The goal set out in this document was to use land and maritime resources more efficiently and effectively, based on the resource endowments and comparative advantage of particular parts of the country, and in response to growing pressures on land resources from industrialization, urbanization, and agriculture in the context of China's rapid economic growth. Environmental protection and ecological development – which had been moving up the national policy agenda – were central drivers. The document classified regions along a spectrum from those that are considered 'optimal' through to 'prohibited' for development, and identified three major types of zone or region: those suitable for urbanization, those for rural or agricultural production, and ecological zones. These were mapped into non-contiguous areas across China's land mass. There was still an underlying east-central-west demarcation: the most developed areas around the Yangzi and Pearl River Deltas and the Bohai rim were considered optimal for development, while there was more 'space' for development in central and western China.

The spatial frameworks that this document sets out were reinforced in the twelfth five-year programme (2011–2015),[85] which included a series of maps showing main agricultural production areas, ecological protection zones, and urban clusters. During this period the impact of urbanization was a particularly important factor for thinking about changes in regional policy. With urban resident population – counting long-term residents, not just those with the full urban household registration or *hukou*[86] – exceeding half of the total population by the time of the twelfth programme, compared to less than one fifth of the population in 1978, the location of major cities became a key driver of economic growth and hence of the spatial structures of regional political economy. This can be seen in the mapping of functional zones, in which major urban clusters – rather than broad geographical descriptors such as 'coastal regions' – were at the forefront of areas for 'optimal development'. This structure reflects a global trend for urban centres to lead job creation and economic development, precisely the shift from surfaces to global networks of urban clusters that has been a wider feature of this era of globalization. It was developed further in material on urban clusters in the thirteenth five-year programme and supported by the continued policy priority given to the upgrading of China's railway network by 2020 and beyond.

Reflecting this linkage between urban development and regional policy, the structure of the twelfth five-year programme brought together material on urbanization with other elements of regional policy.[87] But the conceptual shift from surfaces to networks of urban clusters was only partial. The programme's section on regional policy (section 5) begins with strategy on comprehensive regional development, then the implementation of the strategy of functional zones, followed by pushing forward with urbanization. Its focus on balanced development was largely inherited from the previous phase of policy making around four macro-regions. Here, Develop the West was cited as the 'highest priority' in regional policy, alongside continued injunctions to develop northeast and central China, strengthen support for the 'old revolutionary, ethnic minority, border, and poverty-stricken' regions of the country, and a goal that the eastern coastal regions would continue to play a leading role in China's overall economic development.

This leading role for the coastal regions was particularly relevant in the context in which Chinese policy makers found themselves at the beginning of the second decade of the twenty-first century. In the aftermath of the significant challenge posed by the global financial crisis to China's export industries, which were concentrated in certain coastal provinces (particularly Guangdong, Jiangsu, and Zhejiang), long-standing policy goals of moving up the value chain and increasing competitiveness were re-emphasised. In regional terms, this primarily meant a focus on adding value to industries in parts of coastal China where development had caught up most with other Asian economies, in particular in centres such as Shenzhen and Shanghai.

As well as the section on regional policy, other relevant material in the twelfth five-year programme was to be found in section 12 on 'opening up' China's economy. This dealt with the ways that various regions should enhance their 'openness', using a different regional demarcation: the coastal regions should deepen their openness to the rest of the world as they restructure and move to higher value-added economic activity; the inland regions should attract domestic and international companies to move operations inland, including through the establishment of trade processing bases (in other words, a goal of integrating into regional production networks in the way that coastal regions had done previously); and (inland) border regions should use border trade zones and the development of transborder infrastructure linkages to take advantage of their proximity to neighbouring territories (Tibet is notably not listed among border regions in this context, suggesting that opening Tibet to its land neighbours was not part of the policy thinking in 2011). This latter point is the 'bridging' function for western China highlighted in the launch of the 'new round' of Develop the West in 2010. By this stage, capital for development projects in western China was no longer scarce.

In addition, there were other areas of the programme that more or less explicitly had a regional dimension to them. One such area of growing strategic importance with global dimensions was energy security. Here the programme mentions a plan to construct five domestic regional energy bases – to be located in Shanxi, the Ordos basin in Inner Mongolia, the eastern Inner Mongolia Autonomous Region, southwest China, and Xinjiang – and a coastal nuclear belt.[88] Another aspect of energy strategy is the development of renewable energy (usually defined in China to cover solar and wind power, hydropower, nuclear, and biomass), which has regional relevance given that solar, wind and hydropower resources are all concentrated in regions away from the coast. Hydropower resources are richest in southwest China, as the aggressive building of dams in Sichuan and Yunnan over recent years shows. Solar power is also strongest in western regions, particularly Tibet, and the twelfth programme included plans for six major solar energy bases in western China. Wind resources are most plentiful in Inner Mongolia. But the exploitation of these resources faces substantial challenges for energy grid development as they are located far away from major population and economic centres.

A further regional dimension of energy security has been the development since the mid-2000s of oil and gas pipelines across land into western China, partly to avoid the 'Malacca dilemma' resulting from China's dependence on oil imports, some 80 percent of which travel through the Malacca straits. The five main pipeline routes enter China via its land border provinces, in particular the northeast city of Daqing (the China–Russia oil pipeline), Xinjiang (the China–Kazakhstan oil pipeline and the China–Central Asian gas pipeline), and Yunnan's provincial capital Kunming (the China–Myanmar oil and gas pipelines).[89] This increased the strategic importance of these locations to policy makers and reinforced the idea that the western regions in particular should play a role in opening up to the west of China, as the strategic importance of Eurasia to China has grown.[90] These ideas were to be developed further after 2013 in the belt and road initiative, though the relative importance given to energy appears to have declined somewhat.

Regional policy to 2020 and beyond

The twelfth programme was most likely heavily influenced by the thinking of Xi Jinping, who chaired its drafting committee as vice president, and Li Keqiang, then executive vice premier. After the Party's leadership transition in November 2012 in which Xi took the top party job, followed by government transition in March 2013 when Li became premier, a growing impact of the new leadership began to be felt on regional policy. This moves beyond the ideas and frameworks set out in the twelfth programme to include new ideas, which fall into three broad areas. This period has also seen the upgrading to national policy of numerous regional policies devised at the provincial level, many of which have been designed to encourage cooperation in economic development across administrative boundaries, such as the Shanxi–Shaanxi–Henan Golden Delta Regional Cooperation Plan.[91]

The first new area consists of three new regional initiatives, namely the belt and road initiative, integration of the region around Beijing–Tianjin–Hebei, and the Yangzi river economic belt. The first initiative – which combines the development of a Eurasian 'silk road economic belt' and a '21st-century maritime silk road' – has a clear global scope, though one motivation for the belt part of it (less so for the road) appears to be as a way of stimulating further economic development and openness in some of China's less developed and less open inland provinces, and it can be seen as bringing together Develop the West and China's 'going out' strategy for outward investment as a 'successor to both'.[92]

The other two initiatives relate primarily to developments within China. Beijing–Tianjin–Hebei regional integration covers these three provincial units, with the goal of improving the economic and spatial structures, and enhancing economic efficiency and effectiveness. Dealing with the congestion of the capital is a major goal, with 'non-capital functions' to be moved out to Hebei where possible, and the residential population density in the central city area reduced. In April 2017, the central authorities developed this further by establishing Xiong'an New Area in Hebei province about 100 kilometres southwest of Beijing. Plans for urban construction in the area are designed to facilitate the absorption of functions from the more congested parts of this region.[93]

The Yangzi River economic belt has as its major priorities engaging in ecological 'green' development and developing a strong transport and logistics corridor along a collection of provinces stretching from the Yangzi River Delta inland to western

China. Three major urban clusters – Yangzi River Delta, mid reaches of the Yangzi (around Wuhan), and Chengdu–Chongqing – will play a central role, with Chongqing also highlighted as playing a role as a pivot and hub. The policy document talks about tapping domestic demand and narrowing development gaps between coastal, inland and western China. This initiative was first indicated in December 2012, one month after the national party congress, when premier-in-waiting Li Keqiang toured cities along the Yangzi River. The scale of the two initiatives is markedly different. In 2014 Beijing–Tianjin–Hebei accounted for 8.1 percent of the national population and 9.7 percent of China's GDP, as well as 14.2 percent of national trade (due to the substantial imports to this area, rather than exports),[94] whereas the 11 provinces of the Yangzi River economic belt – Shanghai, Jiangsu, Zhejiang, Anhui, Jiangxi, Hubei, Hunan, Chongqing, Sichuan, Guizhou, and Yunnan – accounted for 42.9 percent of population, 41.6 percent of GDP, and 40.8 percent of total trade.[95]

The second broad area of regional policy innovation, namely the establishment of free trade zones (FTZs), is different in nature, and has a clear external focus. This was a new initiative initially publicized by Li on a visit to Shanghai shortly after he became premier. The China (Shanghai) Pilot Free Trade Zone was formally established in September 2013, with goals of carrying out national policy pilots (hence 'China' in the zone's name); these cover administrative reform and the streamlining of government approvals, facilitating trade and investment to bring China in line with what policy makers referred to as 'high international standards' such as the introduction of a 'negative list' approach to investment approvals,[96] and financial reform, especially around the internationalization of the RMB. Pursuing these goals through limited pilots speaks to the gradual and controlled implementation of the leadership's wider reform agenda, as well as part of a response to the US promotion of the Trans-Pacific Partnership (TPP) from which China was absent.[97] The Shanghai zone was followed by free trade established zones in Guangdong, Fujian and Tianjin in 2015, and in 2017 the establishment of seven further zones in Liaoning, Zhejiang, Henan, Hubei, Chongqing, Sichuan, and Shaanxi.[98]

These zones play a different role in thinking about regional policy and political economy to the elements set out in the twelfth programme. Owing something to the historical experimentation carried out in the SEZs along the south coast, they have been thought of as providing a platform for the more rapid economic and commercial development of these parts of China. But it is probably the other function of the original SEZs – imputed with the benefit of hindsight as much as anything – which is more relevant to understanding the policy relevance of the new FTZs. This is their ability to act as a testing ground for policies that may go national, particularly in areas where the policies being tested are a fairly natural extension of the existing policy frameworks. The exception is the financial sector, where the extent of China's financial globalization is currently limited, and some of the financial reforms in Qianhai, part of Shenzhen, for example, might take that location closer in regulatory terms to Hong Kong than to other Chinese cities.

The third broad new area of regional policy is primarily domestic in focus, but with a global dimension, and concerns further development of government thinking on urbanization and the role of urban clusters as the backbone of the further development of spatial strategy. Details were set out in the promulgation of a programme for a 'new type' of urbanization covering the period from 2014 to 2020.[99] The overarching concept of this plan was that urbanization should support and promote a wide range of national policy goals, from better coordinated regional development, to stimulating domestic demand and economic upgrading, as well as making better use of land and energy resources and improving the environment. A key target is the 'proper' or 'full' urbanization of migrants already resident in urban areas, including through *hukou* (household registration) reform, with an intermediate target that 60 percent of the population should be resident in cities by 2020, with three quarters of these (i.e. 45 percent of the total population) holding urban *hukou*. This links to the ongoing efforts to integrate and coordinate urban and rural development, with a long-term goal of removing the divisive dual urban–rural structure of society in China. Other goals set out in the programme were more effective use of urban land, improved quality of urban construction and better management of urbanization, all of which recognise the problems that many cities in China have been encountering. The document also called for the development of urban clusters in central, western and northeast China, building on those around Beijing, the Yangzi River Delta, and Pearl River Delta – which the government intends to upgrade into a globally-competitive Greater Bay Area.[100] The emphasis on urban clusters reflects a further development of the spatial concepts that had been explored previously in the twelfth five-year programme.

These further evolutions in regional and urbanization policy were reflected in the thirteenth five-year programme (2016–2020), approved by the National People's Congress (NPC) in March 2016.[101] The relevant passages demonstrate the ways that policy making in this area has developed gradually, as well as the new priorities under the post-2012 leadership. There is greater space in the programme given to regional policy than in the twelfth programme, with separate chapters on the 'new type' of urbanization – reflecting the priorities set out in 2014 – and 'coordinated regional development', still the underlying concept for regional policy. While the material on regional development begins with the broad regional strategies for the four macro-regions (described elsewhere in government documents as the 'four regions'), it is followed by section on Beijing–Tianjin–Hebei, the Yangzi River economic belt, regions with 'special characteristics' (a new term to encompass old revolutionary base areas, ethnic minority regions, border regions, and 'difficult regions'), and the maritime or 'blue' economy, which expands significantly on a brief mention of this in the twelfth five-year programme back in 2011. Along with the belt and road initiative – which is mentioned only briefly, but at the start of this section – this shows the priority given to what policy makers had since 2015 called the 'three major initiatives [or strategies]' of regional policy, made clear in both Li Keqiang's and the National Development and Reform Commission's work reports presented to the NPC in March 2016 and March 2017.[102] At the same time, in the manner of Chinese policy documents, which accumulate new initiatives on top of old ones, emphasis is given to the 'main priority' of regional policy still being Develop the West, to plans for a new decade-long strategy to support the 'rise of the central regions' from 2016–2025, continued priority to deal with the legacy problems of the northeast and other old industrial base areas, and bringing out the 'leading role' of the eastern coastal regions in helping the country move to the next stage of development. The FTZs get a brief mention, as do Hong Kong and Macao. The thirteenth programme also looks for further strengthening of linkages between urban clusters, with targets for extending the high-speed rail network, from the 19,000 kilometres goal by 2020 to 30,000 kilometres by 2030, covering more than 80 percent of large cities in China. The most recent statement of Party strategy, Xi Jinping's speech to the 19th National Congress of the Party in October 2017, suggests that these regional policy priorities will continue for some time.

A number of themes emerge from this review of regional policy since the establishment of the PRC. First, these most recent statements of regional policy show that, from a geographical perspective, regional policy has increasingly utilised constructs that cross provincial boundaries and move beyond the earlier demarcation of China into three and then four macro-regions, as indicated by each of the 'three initiatives'. A major goal of regional policy is therefore to counter the natural tendency to the formation of 'administrative region economy', whereby administrative boundaries and competition for resources constrain cooperation and coordination across administrative units.[103] This review also shows the importance of agenda setting in the development of policy,[104] and the role that provincial and other sub-national official and semi-official actors have played in this process, though the degree of provincial agency remains a subject of debate. It further highlights the often contested and sometimes fragmented nature of policy making, which involves a growing range of bureaucratic actors, as well as the way that policy develops gradually over time.

Another theme is the growing need to use regional policy to ensure China's continued development at a national level, whether through pilot programmes, by exhorting the coastal regions to lead this process, or other urban clusters acting as growth poles. This is a reflection of the scale and importance of many of China's regions as the overall economy has grown so rapidly in size. Finally, there is an implicit acknowledgement of the ongoing challenges facing policy makers as a result of diverse and uneven development across regions, and the limited success to date that regional policy has had in addressing that. With this in mind, it is worth looking at some outline economic data on China's regions.

Uneven development and the global interactions of China's regions

The challenge of finding optimal balance in development across China's regions is a response to uneven, unbalanced regional development in China. But the extent and nature of this uneven development is difficult to gauge for policy makers and academics, and depends on what is measured; for example, the World Bank has noted that the biggest differentials may be between urban and rural areas rather than between regions,[105] and pursuing development based on urban clusters may not address this effectively. This chapter concludes with a brief review of some of the recent data on regional economic development, which makes clear the sort of divergent regional trajectories with which policy makers have to deal.[106]

Part of the challenge in mapping regional political economy is that data is collected in line with the administrative structures on which governance is based. At the sub-national level, a cursory examination of some data points at the provincial level gives an idea of the changing nature of regional development. In 2000, the seven provincial units with the fastest annual GDP growth were all in eastern China, namely Zhejiang, Beijing, Guangdong, Shanghai, Tianjin, Jiangsu, and Shandong, all of which recorded provincial GDP growth above 10 percent. By 2010, eight of the ten fastest growing provinces by the same measure were in central or western regions, namely Chongqing, Qinghai, Sichuan, Inner Mongolia, Hubei, Anhui, Hunan, and Shaanxi, all with GDP growth rates of 14.5 percent or more. Only Tianjin (the coastal port city near Beijing) was among the top group in both 2000 and 2010, its continued rapid growth perhaps the result of investment that followed political connections in the central leadership. The other province in the top ten in 2010 was Hainan, which has consistently had its own development path determined by a range of factors not necessarily driven by its regional location.[107]

Reflecting this shift, aggregating GDP growth data across eastern, central and western regions suggests that the fastest overall growth moved from the coast to the inland regions from around 2007. This may partly be a result of policies such as Develop the West, though the impact of such policies is difficult to measure given their broad 'omnibus' nature. Other likely reasons for this shift are connected to China's relationships with the global economy. First, the economic benefits to China from the commodity boom in the 2000s disproportionately accrued in inland provinces where mineral resources are more substantial.[108] This shows up in faster growth in fixed asset investment in inland China, though in 2009 for example eastern China still accounted for a greater share of such investment on a population weighted basis. Second, the global economic crisis had a greater immediate impact on the coastal regions, because they were more deeply integrated into the global economy than those in inland China, though as the case of the Pearl River Delta will demonstrate, parts of coastal China used this as a catalyst to adjust their economic model and move up the value chain.

Even the clear consensus that regional development gaps had grown during the 1990s and since has since been challenged. Recent studies have shown that when per capita indices are calculated using resident rather than registered populations, various measures of regional difference show declines in inter-provincial and coastal-inland inequality since the mid-2000s.[109] And – partly reflecting the urban-rural divide mentioned above – inequality may be a feature within regions, as

34 A historical overview

much as between them; Zhang Weiwei comments that '20 percent of the areas within the regions such as Chongqing, Chengdu and Xi'an are creating 80 percent of the wealth in interior China'.[110]

From a global perspective, the coastal regions remain much more closely integrated into the global economy, as demonstrated by provincial trade statistics. In 2014, coastal provinces – which provide just over half of national GDP – accounted for well over 80 percent of China's total trade.[111] Just three provinces – Guangdong, Jiangsu, and Zhejiang – accounted for almost half of total trade, and over half of total exports (if Shanghai and Shandong are included, then the proportion of total trade is closer to two thirds).[112] These three provinces are effectively the regional base for China's trade surplus, and the surplus of Guangdong, Jiangsu and Zhejiang taken together is in the region of 140 percent of China's total trade surplus; in other words, the rest of China runs an external trade deficit (this does not take account of domestic trade between provinces). Although coastal provinces dominate, the trade volumes of central and western provinces have generally been growing faster than on the coast over recent years, and coastal dominance was even greater in the past: in 2009, nine coastal provinces accounted for over 90 percent of China's total trade, with Guangdong alone accounting for 28 percent. However, the coast's earlier dominance as a destination for foreign direct investment has reduced in scale: in the 1990s the coast was the location for over 90 percent of utilized FDI, but for 2011–2013 that proportion had declined to two thirds (Guangdong and Jiangsu together accounting for nearly one quarter of the national total).[113]

As the impact of the financial crisis and China's credit-fuelled response to it has gradually been digested, more complex and varied regional growth patterns have emerged, featuring growing divergence in GDP growth rates between provinces. In early 2015, the less developed and more investment-reliant provinces were in general displaying weakest economic growth, while there has been much discussion about the challenges facing northeast China, resulting in a series of new central government policy measures through early 2017, involving support from developed coastal provinces. The challenge of maintaining and balancing economic growth across China's regions will continue. We now turn to look at the way some of these issues have played out at a local level, starting with the Pearl River Delta in southern China.

Notes

1 Note that the 15 deputy provincial level cities have enhanced administrative powers but their administrative territorial rank rests with the prefecture level. Carolyn Cartier. 'A political economy of rank: the territorial administrative hierarchy and leadership mobility in urban China'. *Journal of Contemporary China*, 25: 100, p. 533.
2 Liu Junde. 'Regional cooperation in China's administrative region economy: its links with Pan-Pearl River Delta development'. In Xu Jiang and Anthony Yeh, eds. *China's Pan-Pearl River Delta: regional cooperation and development*. Hong Kong: Hong Kong University Press, 2011, pp. 63–78.

3 This is reflected in the idea of 'natural economic territories', though the historical role of political, social, and economic forces should lead us to question the idea that these are 'natural'.
4 Tim Oakes. 'China's provincial identities: reviving regionalism and reinventing "Chineseness"'. *The Journal of Asian Studies*, 59: 3, 2000, p. 684.
5 David S. G. Goodman. 'The politics of regionalism: economic development, conflict and negotiation'. In David Goodman and Gerald Segal, eds. *China deconstructs: Politics, trade and regionalism* (London: Routledge, 1994), p. 3.
6 Hans Hendrischke and Feng Chongyi, *The political economy of China's provinces: Comparative and competitive advantage* (London: Routledge, 1999), p. 2. See also Yang Long, 'China's regional development policy', in John A. Donaldson, ed. *Assessing the balance of power in central-local relations in China*. London and New York: Routledge, 2016, p. 78.
7 This is akin to the process of 'institutionalization of regions', under which 'the state stabilizes and adjusts scale – global, national, regional, and local – as a strategy of accumulation'. Carolyn Cartier. 'Origins and evolution of a geographical idea: the macroregion in China.' *Modern China*, 28: 1, 2002, pp. 79–143.
8 Cartier, 'Origins and Evolution' (see note 7), p. 128.
9 G. William Skinner, ed. *The city in late imperial China*. Stanford, CA: Stanford University Press, 1977. The nine are Manchuria, north China, northwest China, lower Yangzi, middle Yangzi, upper Yangzi, southeast coast, Lingnan (Guangdong/Guangxi) and Yungui (Yunnan/Guizhou). More recent 'standard accounts' identify eight regions: northeast China, north China, central China, south China, southwest China, inner Mongolia, northwest China, and Qinghai–Tibet; cited in Cartier, 'Origins and Evolution' (see note 7), p. 83. Skinner's model has been criticized on both empirical and theoretical grounds. One line of criticism is that the model does not take sufficient account of transregional trade (Cartier, 'Origins and Evolution', pp. 117–127), and in respect of the Yungui macro-region, for example, that it 'does not pay appropriate attention to cross-regional connections between Yunnan and Southeast Asia and beyond'; Yang Bin. *Between winds and clouds: the making of Yunnan (second century BCE to twentieth century CE)*. New York: Columbia University Press, 2009, p. 230.
10 The 'principal-agent' approach. For example, see: Zhong Yang. *Local government and politics in China: challenges from below*. New York: M.E. Sharpe, 2003.
11 Kjeld Erik Brødsgaard. *Hainan: state, society, and business in a Chinese province*. London and New York: Routledge, 2009, p. 81.
12 For a description of how this works in the telecoms sector, see: Yeo Yukyung. 'Between owner and regulator: governing the business of China's telecommunications service industry'. *The China Quarterly*, 200, p. 1019.
13 Meg E. Rithmire. 'China's 'New Regionalism': subnational analysis in Chinese political economy'. *World Politics*, 66: 1, 2014, p. 189.
14 Alvin So and Yin-wah Chu. *The global rise of China*. Cambridge, UK: Polity, 2016.
15 Rithmire, 'China's "New Regionalism"' (see note 13), pp. 173–174; So and Chu, *The global rise of China* (see note 14), p. 132.
16 National Development and Reform Commission. 'Report on the implementation of the 2015 plan for national economic and social development and on the 2016 draft plan for national economic and social development', delivered at the Fourth Session of the Twelfth National People's Congress, 5 March 2016, http://news.xinhuanet.com/english/china/2016-03/18/c_135202412_3.htm, emphasis added.
17 Kjeld Allan Larsen. *Regional policy of China, 1949–85*. Manila: Journal of Contemporary Asia Publishers, 1992; David Goodman and Gerald Segal, eds. *China deconstructs: politics, trade and regionalism*. London: Routledge, 1994; David Goodman, ed, *China's provinces in reform: class, community and political culture*. London and New York: Routledge, 1997; Hendrischke and Feng, *The political economy of China's provinces* (see note 6); Jae Ho Chung, ed. *Cities in China: recipes for economic development in the*

36 A historical overview

reform era. London and New York: Routledge, 1999; John Fitzgerald, ed. *Rethinking China's provinces*. London: Routledge, 2002; Zheng Yongnian. *De facto federalism in China: reforms and dynamics of central-local relations*. Hackensack, NJ: World Scientific, 2007; Kjeld Erik Brødsgaard, *Hainan* (see note 11); John A. Donaldson. *Small works: poverty and economic development in southwestern China*. New York and London: Cornell University Press, 2011; Tim Summers, *Yunnan: a Chinese bridgehead to Asia: a case study of China's political and economic relations with its neighbours*. Oxford, UK: Chandos, 2013.

18 This literature is vast. For a good example see: Stevan Harrell. *Ways of being ethnic in Southwest China*. Seattle, WA and London: University of Washington Press, 2001.
19 Robert A. Kapp. *Szechwan and the Chinese republic: provincial militarism and central power, 1911–38*. Yale, CT: Yale University Press, 1974; Chris Bramall. *In praise of Maoist economic planning: living standards and economic development in Sichuan since 1931*. Oxford, UK: Clarendon Press, 1993; David Goodman. *Centre and province in the People's Republic of China: Sichuan and Guizhou, 1955–1965*. Cambridge, UK: Cambridge University Press, 1986.
20 For example: Ben Hillman and Gray Tuttle, eds. *Ethnic conflict and protest in Tibet and Xinjiang*. New York: Columbia University Press, 2016; Michael E. Clarke and Douglas Smith, eds. *China's Frontier regions: ethnicity, economic integration and foreign relations*. London and New York: I. B. Tauris, 2016; Han Enze. *Contestation and adaptation: the politics of national identity in China*. Oxford, UK: Oxford University Press, 2013; Colin Mackerras and Michael Clarke, eds. *China, Xinjiang and Central Asia: history, transition and crossborder interaction into the 21st century*. London and New York: Routledge, 2009; Gabriel Lafitte. *Spoiling Tibet: China and resource nationalism on the roof of the world*. London: Zed Books, 2013; Michael Dillon. *Xinjiang: China's Muslim far northwest*. London and New York: Routledge, 2004.
21 Peter T. Y. Cheung and James T. H. Tang. 'The external relations of China's provinces'. In David Lampton, ed. *The making of Chinese foreign and security policy 1978–2000*. Stanford, CA: Stanford University Press, 2001; Linda Jakobson and Dean Knox. 'New foreign policy actors in China', SIPRI Policy Paper 26, September 2010, www.lowyinstitute.org/publications/new-foreign-policy-actors-china.
22 Li Mingjiang. 'Central-local interactions in foreign affairs'. In John A. Donaldson, ed. *Assessing the balance of power in central-local relations in China*. London and New York: Routledge, 2016.
23 Cheung and Tang, 'The external relations of China's provinces' (see note 21).
24 Summers, *Yunnan* (see note 17).
25 Jacobson and Knox, 'New foreign policy actors' (see note 21); Chen Zhimin. 'Coastal provinces and China's foreign policy-making', 2013, www.cewp.fudan.edu.cn/attachments/article/68/Chen%20Zhimin,%20Coastal%20Provinces%20and%20China%27s%20Foreign%20Policy%20Making.pdf.
26 Shahar Hameiri and Lee Jones. 'Rising powers and state transformation: the case of China'. *European Journal of International Relations*, 21: 1, 2016, p. 84.
27 Chen Xiangming. *As borders bend: transnational spaces on the Pacific Rim*. Boulder, CO: Rowman and Littlefield, 2005.
28 Dorothy Solinger. *Regional government and political integration in southwest China 1949–1954*. Berkeley, CA: University of California Press, 1977.
29 Thomas Mullaney. *Coming to terms with the nation: ethnic classification in modern China*. Berkeley, CA: University of California Press, 2011; Louisa Schein. *Minority rules: the Miao and the feminine in China's cultural politics*. Durham, UK and London: Duke University Press, 2000, pp. 96–97.
30 Justin Yifu Lin and Peilin Liu. 'Development strategies and regional income disparities in China'. In Guanghua Wan, ed. *Inequality and growth in modern China*. Oxford, UK: Oxford University Press, 2008, p. 59. See also: Nicolaas Groenewold, Anping Chen and Guoping Lee. *Linkages between China's regions: measurement and policy*. Cheltenham, UK: Edward Elgar, 2008, pp. 31–33.

31 The dominant coastal-inland demarcation of China's geography for much of the 20th century had its 'origins in the imperialist penetration of the mid-nineteenth century'; Larsen, *Regional policy* (see note 17), p. 87.
32 Larsen, *Regional policy* (see note 17), pp. 110–114.
33 Mao Zedong, 'On the ten great relationships', 1956. A later version of the text, from which somewhat different emphases can be deduced, replaced 'more than 90 percent' with the 'greater part'; Stuart Schram. 'Chairman Hua edits Mao's literary heritage: "On the 10 great relationships" '. *The China Quarterly*, 69, 1977, p. 131.
34 Xu Jiang and Anthony Yeh. 'Political economy of regional cooperation in the Pan-Pearl River Delta'. In Anthony G. O. Yeh, and Xu Jiang, eds. *China's Pan-Pearl River Delta: regional cooperation and development*. Hong Kong: Hong Kong University Press, 2011, p. 105.
35 Larsen, *Regional policy* (see note 17), p. 118.
36 Barry Naughton. 'The Third Front: defence industrialization in the Chinese interior'. *The China Quarterly*, 115, 1988, p. 369.
37 Naughton, 'The Third Front' (see note 36), pp. 365–368.
38 Lin and Liu, 'Development strategies' (see note 30), p. 60. It is worth noting that the original draft of the third five-year plan produced in 1963–1964 had envisaged 70 percent of investment in coastal China, 20 percent in inland areas, and the rest in intermediate areas, until the Third Front intervened; see Naughton, 'The Third Front' (see note 36), p. 352.
39 Maurice Meisner. *Mao's China and after*. New York: Free Press, 1999, pp. 376–386.
40 Naughton, 'The Third Front' (see note 36), pp. 363–364.
41 Naughton, 'The Third Front' (see note 36), p. 351.
42 Covell Meyskens. 'Third Front railroads and industrial modernity in late Maoist China'. *Twentieth-Century China*, 40: 3, 2015, pp. 238–260.
43 Naughton, 'The Third Front' (see note 36), p. 364.
44 Lin and Liu, 'Development strategies' (see note 30), p. 61; Groenewold, Chen and Lee, *Linkages between China's regions* (see note 30), pp. 38–40.
45 David S. G. Goodman. 'The politics of regionalism: economic development, conflict and negotiation'. In David Goodman and Gerald Segal, eds. *China deconstructs: politics, trade and regionalism*. London: Routledge, 1994, p. 2.
46 Lin and Liu, 'Development strategies' (see note 30), p. 61.
47 Larsen, *Regional policy* (see note 17), pp. 78–79.
48 Goodman, 'The politics of regionalism' (see note 45), p. 1.
49 Cited in Lin and Liu, 'Development strategies' (see note 30), p. 61, footnote 7, emphasis added.
50 Larsen, *Regional policy* (see note 17), pp. 137–138.
51 Groenewold, Chen and Lee, *Linkages between China's regions* (see note 30), p. 42.
52 Brødsgaard, *Hainan* (see note 11), p. 33.
53 Larsen, *Regional policy* (see note 17), p. 192.
54 Brødsgaard, *Hainan* (see note 11), p. 33.
55 Dali Yang. 'Patterns of China's Regional Development Strategy'. *The China Quarterly*, 122, 1990, p. 248. Enright has 91 percent of foreign direct investment going to coastal provinces from 1979–1990; Michael Enright. *Developing China: the remarkable impact of foreign direct investment*. London and New York: Routledge, 2016, p. 26.
56 Dali Yang. 'China adjusts to the world economy: the political economy of China's coastal development strategy'. *Pacific Affairs*, 64: 1, 1991, p. 42. The full name was 'outward-oriented development strategy in the coastal areas' (*yanhai diqu waixiangxing fazhan zhanlue*).
57 Yang, 'China adjusts' (see note 56), p. 51. See Chapter 4 of the present book for a subsequent innovative development of this concept.
58 Meisner, *Mao's China* (see note 39), pp. 488–491.
59 Fuh-wen Tzeng. 'The political economy of China's coastal development strategy'. *Asian Survey*, 31: 3, 1991, pp. 270–284.

38 A historical overview

60 Groenewold, Chen and Lee, *Linkages between China's regions* (see note 30), p. 9.
61 Yang, 'Patterns of China's regional development strategy' (see note 55), p. 242.
62 Jiang Xu and Anthony Yeh. 'Political economy of regional cooperation in the Pan-Pearl River Delta' (see note 34).
63 Zheng, *De facto federalism in China* (see note 17).
64 Tian Xiaowen. 'Deng Xiaoping's *Nanxun*: impact on China's regional development'. In John Wong and Zheng Yongnian, eds. *The Nanxun legacy and China's development in the post-Deng era*. Singapore; Singapore University and Press and World Scientific, 2001, pp. 75–94.
65 Carolyn Cartier. *Globalizing south China*. Oxford, UK: Blackwell, 2001, p. 6.
66 Kam Wing Chan. 'China: internal migration'. In Immanuel Ness, ed. *The encyclopedia of global human migration*. Hoboken, NJ: Blackwell, 2013. http://onlinelibrary.wiley.com/doi/10.1002/9781444351071.wbeghm124/abstract.
67 Gerald Segal. 'The Muddle Kingdom? China's changing shape'. *Foreign Affairs*, 73: 3, 1994, p. 57. Goodman and Segal's influential volume of the same year, *China deconstructs*, explores the external economic linkages in detail, though its overall conclusion is a more cautious one than that suggested by the book's title.
68 Published in English as: Wang Shaoguang and Hu Angang. *The political economy of uneven development: the case of China*. New York: M. E. Sharpe, 1999.
69 Yang, 'China adjusts' (see note 56), p. 53.
70 Susan Shirk. *The political logic of economic reform in China*. Berkeley, CA: University of California Press, 1993. Yang, 'Patterns of China's regional development strategy' (see note 55), p. 254.
71 Summers, *Yunnan* (see note 17), pp. 54–58.
72 Brødsgaard, *Hainan* (see note 11), p. 115.
73 Tian, 'Deng Xiaoping's *Nanxun*' (see note 64), p. 78.
74 Groenewold, Chen and Lee, *Linkages between China's regions* (see note 30), pp. 47–49; they date the shift to the eighth five-year plan. See also Xu and Yeh, 'Political economy of regional cooperation' (see note 34), p. 107. The seven regions were the Yangzi River Delta and River Basin, Bohai Rim, Southeast Coastal Region (Guangdong and Fujian), South and Southwest Region, Northeast Region, Five Provinces of Central China, and Northwest Region.
75 Lin and Liu, 'Development strategies' (see note 30), p. 62.
76 Groenewold, Chen, and *Lee. Linkages between China's regions* (see note 30), p. 28.
77 David Goodman. 'The campaign to "open up the west": national, provincial-level and local perspectives'. *The China Quarterly*, 178, 2004; Tim Summers. 'China's western regions 2020: their national and global implications'. In Kerry Brown, ed. *China 2020: what policy makers need to know about the new rising power in the coming decade*. Oxford, UK: Chandos, 2011, pp. 137–171; Gregory T. Chin. 'The politics of China's western development initiative'. In Ding Lu and William A. W. Neilson, eds. *China's western region development: domestic strategies and global implications*. Singapore: World Scientific, 2004.
78 Elena V. Barabantseva. 'Development as localization: ethnic minorities in China's official discourse on the Western Development Project'. *Critical Asian Studies*, 41: 2, 2009; Carolyn Cartier, 'Uneven development and the time/space economy'. In Wanning Sun and Yingjie Guo, eds. *Unequal China: the political economy and cultural politics of inequality*. New York and Oxford, UK: Routledge, 2013, pp. 77–90.
79 Jae Ho Chung, Hongyi Lai, and Jang-Hwan Joo. 'Assessing the "Revive the Northeast" programme'. *The China Quarterly*, 197, 2009.
80 Cited in Hongyi Lai. 'Developing Central China: a new regional programme'. *China: An International Journal*, 5: 1, 2007, p. 119.
81 Lai, 'Developing Central China' (see note 80), p. 109.
82 Tim Summers. 'China's western regions 2020' (see note 77), p. 149.

83 Yang, Long. 'China's regional development policy'. In John A. Donaldson, ed. *Assessing the balance of power in central-local relations in China*. London and New York: Routledge, 2016, pp. 76–104, p. 77.
84 State Council, Guowuyuan guanyu yinfa quanguo zhuti gongnengqu guihua de tongzhi (Notification of State Council plan on national functional regions), 2010, www.gov.cn/zwgk/2011-06/08/content_1879180.htm.
85 From the 11th five-year programme (2006–2010), the name changed from 'plan' (*jihua*) to 'guideline programme' (*guihua*); see: Robert Ash, Robin Porter, and Tim Summers. 'Rebalancing towards a sustainable future: China's twelfth five-year programme'. In Kerry Brown, ed. *China and the EU in context: insights for business and investors*. London: Palgrave MacMillan, 2014, p. 83.
86 Under the *hukou* system, established in the 1950s, each member of the population has a residence permit for a particular location and is classified as either rural or non-rural. Following mass migration in the 1990s and 2000s there are over 260 million migrant workers across China, many of whom hold rural *hukou* but are resident in cities. This means they do not usually have access to the full range of urban public services.
87 Ash et al., 'Rebalancing towards a sustainable future' (see note 85), pp. 102–105.
88 Gangyao. 'Zhonghua Renmin Gongheguo Guomin Jingji he Shehui Fazhan de Shi'er ge Wunian Guihua Gangyao' ('Outline of the 12th five-year programme for national economic and social development of the People's Republic of China'). Beijing: People's Press, 2011, p. 34.
89 Chen Shaofeng. 'Has China's foreign energy quest enhanced its energy security?' *The China Quarterly*, 207, 2011, pp. 600–625.
90 Wang Jisi. 'North, South, East, and West – China is in the "middle": a geostrategic chessboard'. *China International Strategy Review*, 2013.
91 Yang, 'China's regional development policy' (see note 83), p. 93.
92 Emily T. Yeh and Elizabeth Wharton. 'Going West and going out: discourses, migrants, and models in Chinese development'. *Eurasian Geography and Economics*, 57: 3, 2016, p. 288. Chapter 5 of the present book explores this initiative further from a subnational perspective.
93 In administrative terms, this was the 19th of a number of New Areas to be established as elevated administrative areas across numerous cities since 2010, though politically it is by far the most significant.
94 This data incorporates the whole of Hebei, though the focus of the initiative is on parts of the province closer to Beijing and Tianjin.
95 China Statistics Press. *China statistical yearbook 2015*. Beijing: China Statistics Press, 2015, p. 864.
96 The 'negative list' means that foreign investment is permitted in China unless the sector is listed as prohibited or restricted. Previously all sectors were categorized as prohibited, restricted, and permitted.
97 Tim Summers. 'Chinese multilateralism: diluting the TPP'. In Adrian Hearn and Margaret Myers, eds. *The changing currents of transpacific integration: China, the TPP, and beyond*. Boulder, CO and London: Lynne Rienner, 2017.
98 163.com, 'Shangwubu buzhang Gao Hucheng: Guowuyuan jueding xin she 7 ge zimao shixian qu' ('Minister of Commerce Gao Hucheng: State Council decision to establish seven new free trade pilot zones'), 31 August 2016, http://c.m.163.com/news/a/BVQP547F000156PO.html.
99 State Council. 'Guojia xinxing chengzhenhua guihua (2014-2020nian)' ('National programme for a new type of urbanization, 2014–2020'), 17 March 2014, http://politics.people.com.cn/n/2014/0317/c1001-24649809.html.
100 This is explored in Chapter 3 of the present book.
101 *Xinhua*. 'Zhonghuo renmin gongheguo guomin jingji he shehui fazhan de shisan ge wunian guihua gangyao' ('Outline of the 13th five-year programme for national

economic and social development of the People's Republic of China'), 17 March 2016, http://news.xinhuanet.com/politics/2016lh/2016-03/17/c_1118366322.htm.
102 Li Keqiang, 'Report on the Work of the Government', 16 March 2017, http://news.xinhuanet.com/english/china/2017-03/16/c_136134017.htm; National Development and Reform Commission, 'Report on the implementation of the 2015 plan for national economic and social development and on the 2016 draft plan for national economic and social development', delivered at the Fourth Session of the Twelfth National People's Congress, 5 March 2016, http://news.xinhuanet.com/english/china/2016-03/18/c_135202412_3.htm.
103 Liu. 'Regional cooperation in China's administrative region economy' (see note 2).
104 Wang Shaoguang. 'Changing models of China's policy agenda setting'. *Modern China*, 34: 1, 2008.
105 World Bank, China. *From poor areas to poor people: China's evolving poverty reduction agenda – an assessment of poverty and inequality in China*. Washington, DC: The World Bank, 2009, http://documents.worldbank.org/curated/en/2009/03/10444409/china-poor-areas-poor-people-chinas-evolving-poverty-reduction-agenda-assessment-poverty-inequality-china-vol-1-2-main-report.
106 Data for this section is sourced from China Statistical Yearbooks unless otherwise stated. There is a huge amount of data on these questions, and this section only scratches the surface.
107 Brødsgaard, *Hainan* (see note 11).
108 The Economist, 'Rich province, poor province', 1 October 2016, www.economist.com/news/china/21707964-government-struggling-spread-wealth-more-evenly-rich-province-poor-province.
109 John Gibson and Chao Li. 'Rising regional income inequality in China: fact or artifact?' University of Waikato New Zealand, Department of Economics Working Paper in Economics 09/12, 2012. https://ideas.repec.org/p/wai/econwp/12-09.html.
110 Zhang Weiwei. *The China wave: rise of a civilizational state*. Singapore: World Scientific, 2011, p. 45.
111 By location of importers/exporters, the eastern region accounted for 82 percent of total trade and coastal provinces 86 percent; by place of destination or origin in China the figures are 83 percent and 87 percent respectively.
112 Here the distinction between location of importers/exporters and place of destination/origin is greater: for the former, these five provinces account for 64 percent of the total, and for the latter, 70 percent of the total.
113 Enright, *Developing China* (see note 55), p. 26.

3

THE GREATER BAY AREA

From factory floor to global challenger

The part of China that was at the forefront of China's initial engagement with a new intensified phase of globalization from the 1980s was an area of Guangdong province in south China known as the Pearl River Delta (PRD, see Table 3.1). It was no coincidence that this lay adjacent to Hong Kong, then a British colony with strong global economic linkages, giving Hong Kong a central role in this first phase of China's globalization. Today, even without considering the Hong Kong links, the PRD is probably the most globally integrated of China's regions. The latest policy goal is to upgrade this region into a Greater Bay Area, building a globally-competitive advanced economy that will benefit from the comparative advantages of its different components in Guangdong province and the two Special Administrative Regions of Hong Kong and Macao (see Box 3.1). To explore the ongoing dynamic role this region plays within China and in the context of China's globalization, we need to look first at the processes through which the PRD came to prominence as a region of China that would become known as the 'factory of the world', at the responses to the challenges brought by the global financial crisis, and the changing relationship today between the PRD and Hong Kong.

The global emergence of the Pearl River Delta (PRD)

Before the inauguration of 'reform and opening up' in the late 1970s there was little talk of the Pearl River Delta either inside or outside China. Neither was Guangdong province a particular focus of the Chinese economy, though the provincial capital Guangzhou had been one of the earliest Chinese cities that industrialized from the late Qing dynasty onwards, and had featured in the complex political changes during the first part of the 20th century. In 1978, Guangdong was at best a mid-ranking province in terms of GDP per capita. By 2001, the contrast was stark: Guangdong's GDP per capita of RMB13,612 placed it second

BOX 3.1 PEARL RIVER DELTA CONCEPTS

- Pearl River Delta region: the nine municipal administrative jurisdictions of Guangzhou, Shenzhen, Dongguan, Foshan, Huizhou, Jiangmen, Zhuhai, Zhongshan, and Zhaoqing. Total population c.59 million, accounting for around 23 percent of Guangdong province's land area.
- Greater PRD: the Pearl River Delta plus the Hong Kong and Macao special administrative regions (SARs).
- Pan-Pearl River Delta (PPRD): a separate grouping of nine provinces covering most of southern China, plus Hong Kong and Macao, instigated by Guangdong province in 2004.
- Guangdong–Hong Kong–Macao (*Yue-Gang-Ao*) Greater Bay Area: a newer concept referenced in the central government's March 2015 document on the 'belt and road initiative' and in the 13th five-year programme (2016–2020), which has since become the central policy concept for this region.

TABLE 3.1 Pearl River Delta economic and social data (2015)

	Total for PRD	*Proportion of national total*
Land area	54,743 sq. km	0.57%
GDP	RMB 5,597 billion	9.1%
Permanent population	59 million	4.3%
Per capita GDP	RMB 95,279	N/A
Exports	USD 609 billion	26.8%
Actual FDI	USD 25.6 billion	N/A

Source: Hong Kong Trade Development Council, 'PRD Economic Profile', 2 December 2016, www.hktdc.com/business-news/article/Fast-Facts/PRD-Economic-Profile/ff/en/1/1X000000/1X06BW84.htm.

only to Zhejiang among China's provinces, though still behind that of the three leading cities of Shanghai, Beijing, and Tianjin.[1] Today its wealthy economy is larger in US dollar terms than that of Indonesia.

This transformation – a microcosm of which is dramatically revealed in the account of one part of the province, *Chen Village*[2] – followed the initiation of reform and opening up under Deng Xiaoping from December 1978. It was given additional impetus by Deng's 'southern tour' in early 1992, which included Shenzhen and Zhuhai in Guangdong, when he said that Guangdong should catch up with the four Asian dragons (Hong Kong, Singapore, South Korea and Taiwan) in two decades.[3] China's initial opening up under Deng – with his injunction that 'some regions should get rich first' – was not as something that applied across the country, but the experimental opening up of parts of coastal China in a process that was gradually expanded geographically as the 1980s and 1990s rolled on. Guangdong, then under the stewardship of party secretary Xi Zhongxun, father of current Party General Secretary Xi Jinping, was at the heart of the initial experimentation and home to three of the four special economic zones (SEZs) formally established in 1980, namely Shenzhen, Zhuhai and Shantou (the fourth SEZ, Xiamen, lies further up the coast in neighbouring Fujian province).

The location of these SEZs gave Hong Kong a central role in the opening to foreign trade and investment, while Shenzhen's emergence as the most successful of these SEZs – today it has one of the most vibrant economies of any city in China and has been called 'the youngest of world cities'[4] – was directly attributable to its proximity to Hong Kong. The model was 'Hong Kong as the shop window and China as the factory floor' (*qiandian houchang*),[5] whereby Hong Kong capital invested in the PRD for manufacturing and assembly for export – trade processing as part of the development of transnational production networks – while the externally-facing sales and marketing and business services were clustered in Hong Kong. This was driven in the 1980s mainly by small businesses, which 'did not have the capacity to engage in long-distance offshore production', and for whom kinship, ethnic, and community ties (*guanxi*) could be used to facilitate business engagement. Sociologist Alvin So calls this 'informal, societal integration' between Hong Kong and its hinterland, to which the colonial British Hong Kong government took a hands-off attitude.[6] This not only expedited the shift of Hong Kong's increasingly uncompetitive manufacturing into the PRD and the upgrading of Hong Kong's economy to one focusing on services and finance, but led to the emergence of the PRD as a major export-oriented region for manufacturing and assembly, attracting almost half of total foreign direct investment into China in the 1980s.[7] The relationship between Hong Kong and the PRD during this period was one of complementarity and comparative advantage.

It was therefore in the PRD (or parts of it) rather than across the whole province of Guangdong where economic development was focused. The PRD – which accounts for less than one quarter of the province's land mass and half of its population – saw its proportion of Guangdong's GDP rise from 47.7 percent in 1980 to 55.9 percent in 1990 and 76.4 percent in 2000.[8] This had wider implications for national policy makers as they reflected on the lessons of the 1980s, with the conclusion that 'Guangdong's developmental success was seen as proof of the efficacy of export-oriented development'.[9] This should be seen as a bottom-up business-driven process facilitated and encouraged by the 'reform and opening up' framework and the wider changes under economic globalization, as much as the direct result of Chinese government planning. It reflects the nature of globalization's development during this period, with the development of corporate-driven production networks being facilitated by government policy.[10]

One policy contribution to Guangdong's economic development came from favourable fiscal arrangements with the central government after 1980.[11] Unsurprisingly, there was substantial resistance from Guangdong leaders to the efforts over a decade later by then vice-premier Zhu Rongji to renegotiate central-provincial fiscal relations, which eventually led to the agreement in 1994 of a new formula for the sharing of tax revenues between the central government and provinces.[12] This was not the only area of tension between the province and the centre, and Guangdong's 'blockage of a Sino-American deal on trucking' in the 1990s was an example of these central-provincial difficulties.[13] This was during a period when the central government was relatively weak and there were questions over the state's ability to guarantee China's territorial integrity under the challenge of externally-oriented regionalism.[14] The fiscal reforms pushed through by Zhu in 1994 were an important step in the central government's response to these challenges, and contributed to the receding of questions about territorial integrity over subsequent years.

By 1990 the PRD accounted for 13.0 percent of mainland China's exports, and the remainder of the 1990s and the first part of the 2000s saw the continued growth of the PRD's export industries, to the point where the region became known as the 'factory of the world'. In 2001, the year that China joined the World Trade Organisation, the PRD accounted for 34.1 percent of China's total exports,[15] for 88.2 percent of China's exports of fans, 78.8 percent of telephones, 43.6 percent of video recorders, and 34.9 percent of colour televisions; a tenth of the PRD's total exports were sold by Wal-Mart. As well as Hong Kong investors, Taiwanese companies also played an important role in the PRD's manufacturing and export industries, with Dongguan city (north of Shenzhen) a particular favoured location in the PRD, driving the rapid development of Dongguan from a rural township to a city flash with new wealth, though accompanied by exploitation of labour, inequality, and environmental degradation.[16]

The aggregate effect of the PRD's development, and the links between its various cities and Hong Kong, stimulated suggestions from the late 1990s that 'what is emerging is a mega-city of 40–50 million people, connecting Hong Kong, Shenzhen, Guangzhou, Zhuhai, Macau, and small towns in the Pearl River Delta'. In 1995, this metropolitan regional system extended over 50,000 square kilometres, but was characterised by scattered units becoming interconnected functionally through a multimodal transport and logistics system.[17] At that point in time, there were five new airports under construction, as well as new container ports such as Yantian in Shenzhen (east of Hong Kong), which today is one of the largest globally. Such mega-cities are a new spatial form that plays a key role in connecting the global economy as centres of 'economic, technological and social dynamism' and of 'cultural and political innovation', with Hong Kong acting as a 'nodal point' in this region. However, at the same time as concentrating some of the most dynamic developments through global connectivity, mega-cities have a tendency 'internally [to] disconnect local populations that are either functionally unnecessary or socially disruptive'. Mega-cities concentrate the best, but also reflect some of the worst consequences of the integration into a global capitalist economy.[18]

With this in mind, this region can be seen as a number of networked economic areas. There is a dual core to the PRD consisting of Guangzhou and Shenzhen, two very different cities; more recently, policy makers have talked of Guangzhou-Foshan as one urban centre, Shenzhen-Dongguan as a second, and Zhuhai-Zhongshan (on the west bank of the Pearl River) as a third. Around this the idea of a 'greater PRD' has in the past been used to bring in Hong Kong and Macao along with the rest of the PRD, and highlights the crucial historical role played by Hong Kong in the region's development – though this is now being replaced with the concept of a Guangdong–Hong Kong–Macao Greater Bay Area. The PRD's hinterland extends further into other parts of southern China, and in the 2000s, partly under pressure of competition from the Yangzi River Delta and other provinces along China's coast (Jiangsu overtook Guangdong in attracting foreign direct investment in the 2000s), the Guangdong provincial authorities, with Zhang Dejiang as provincial party secretary, began to look for ways of enhancing connectivity to this hinterland. The outcome was the Pan-Pearl River Delta (PPRD) economic cooperation forum, a group of nine provinces across southern China plus the Hong Kong and Macao special administrative regions, which first met in 2004 to discuss economic cooperation and improving logistics linkages inland from the PRD. This is an example of the sort of state-led development of spatial networks for smoother flows of capital and goods that has informed much of China's regional policy, and features in the other regional developments, including Chongqing and the belt and road initiative.

The global crisis and the PRD

Before the global economic and financial crisis hit in 2008, Guangdong province had good reason to consider itself at the forefront of China's economic development. The provincial economy was the largest in the country, accounting for over one tenth of the national total, though in GDP per capita terms it ranked lower than several other provincial units. Its vanguard status was more to do with its leading role in developing China's much-vaunted export-led growth, a substantial part of which was through the processing trade that linked it into networks of global capital. In terms of foreign trade, Guangdong clearly ranked top among China's provinces, but when the crisis hit in 2008, the province was the part of China most affected by the sudden drop in demand for exports that followed, exposing how dependent the PRD was on labour-intensive and relatively low value added export industries. These were also the reason for the presence in Guangdong (mainly in the PRD) of some 30 million migrant workers from other parts of the country, adding to the challenges of the downturn, not least because of the challenges for public service provision of large migrant populations.

In 2009, Guangdong still accounted for 28 percent of national trade and its total trade to GDP ratio was 107 percent, well above the 59 percent for Jiangsu and 56 percent for Zhejiang. The external nature of the economy can also be seen by the relatively limited role that fixed asset investment has played in the economy – in 2010, the ratio of fixed asset investment to GDP was the lowest among all provinces apart from Shanghai's (the total value of fixed asset investment was ranked sixth).[19] The province's trade volumes rose again by 26.3 percent in 2010, but in the aftermath of the crisis, Guangdong's economy began to go through some major structural shifts. Numerous Hong Kong businesses in the PRD closed down – a reduction from 56,000 to under 40,000 according to data from the Hong Kong Federation of Industries.[20]

When the global financial crisis hit, official thinking had already begun to address the structural challenges facing Guangdong and its PRD. In December 2008 the central government's strategic planning ministry, the National Development and Reform Commission, published an authoritative 'outline plan for the reform and development of the Pearl River Delta (2008–2020)'. The plan suggested that the PRD was underdeveloped and focused on the need for the PRD to reform and integrate internally to maintain its competitiveness. It covered the PRD rather than the Greater PRD (which includes Hong Kong and Macao), though it did talk about areas for cooperation with the two special administrative regions. Targets for both 2012 and 2020 (see Table 3.2) reflected some important national priorities, such as developing the services sector and innovation, but also the relative weakness of the PRD in these areas.

All of these targets reflected emerging themes in national policy to improve the quality and global status of China's economy. The targets themselves were ambitious: against a target of 53 percent for 2012, Guangdong's service sector accounted for under 46 percent of GDP in 2009 (though well over half of GDP in Shenzhen), while exports of services were only 6 percent of total provincial foreign trade, compared to the target set for 2012 of 20 percent. The target for spending on research and development to reach 2.5 percent of GDP in 2012 was particularly ambitious given that Guangdong province's research and development spend in 2008 was only 1.15 percent of GDP, lower than the national average. This ambition was even more evident as the target for Guangdong was set above the aggregate national goal, which in 2011 was to reach 2.2 percent of GDP 2015. The PRD plan also aimed to improve coordination between the nine cities in the PRD, addressing some of the issues of disparities highlighted above, as well as the consequences of 'competitive liberalization' between cities in attracting

TABLE 3.2 Pearl River Delta economic targets

	2012 target	2020 target
GDP per capita	RMB80,000	RMB135,000
Urbanization ratio	80%	85%
Service sector as proportion of GDP	53%	60%
Services trade as proportion of provincial trade	20%	40%
Spending on research and development as proportion of GDP	2.5%	none

Source: National Development and Reform Commission (2008). 'Outline plan for the reform and development of the Pearl River Delta (2008–2020)'.

investment.[21] An example of the provincial government's efforts to even out development to some extent was the 2008 decision to site major steel and petrochemicals projects in Zhanjiang, at the province's western periphery outside the PRD.

Leadership matters, and Guangdong's subsequent development has much to do with the fact that the provincial leadership saw the global financial crisis as an opportunity as much as a threat.[22] A new party secretary, Wang Yang, had taken up his post in late 2007 after serving two and a half years as party secretary in Chongqing. Wang was seen as something of a rising star in Chinese politics, and had served in the State Council (central government) and influential central government strategic economic commission, the National Development and Reform Commission, before moving to Chongqing in early 2005. He placed more emphasis on addressing some of the inequalities in the province, and paid some high-profile early visits to peripheral parts of Guangdong, such as the northern parts of Qingyuan (see below). Provincial policy reflected the evolving national priorities by emphasizing both improving livelihoods and upgrading economic development, through a slogan of 'speeding up [economic] transformation and upgrading, building a contented [*xingfu*] Guangdong'. Wang also promoted experimental reforms in administration.

An article by Wang in the *Southern Daily* newspaper of 18 March 2011 gave insights into his thinking. In it, Wang stressed that economic development should be a means to a contented population, not an end in itself, and legitimized this point by arguing that this was the intention of the Party in making economic development the 'central task' in the reforms following December 1978. However, he argued that both economic development and contentment (*xingfu*) needed to be taken together: the latter could not be achieved without the former. This he compared to making and sharing a cake: a big and good cake was needed before it could be shared equitably. As for the nature of *xingfu*, Wang argued that this had changed over time, from feeding the population in the early years of the PRC, to the ability to earn money from the 1980s, to a wider range of demands in the early 21st century, encompassing better income distribution, social insurance, health and education.[23] He stressed that this was a long-term project, which required sustainable economic development and improvement of livelihoods. Although Wang's approach in Guangdong was often characterized as a 'Guangdong model', these messages were in line with national policy statements, though delivered in an open and engaging style that was not the norm across Chinese officialdom.

These themes were reflected in the management of the PRD's workforce. Minimum wages in Guangdong rose an average of 21.1 percent in 2010 and 18.5 percent in 2011, varying from RMB1300/1320 per month in Guangzhou and Shenzhen respectively – the highest in China at the time – down to RMB850 in peripheral cities. Subsequent rises were higher than the 13 percent annual rise mandated in the twelfth five-year programme (the national policy document setting out economic and social policy goals for 2011–2015), though the minimum wage has become less relevant as market forces and demographics have interacted to push up factory wages; in 2011, anecdotal evidence suggested that these tended to be at least RMB2000 per month, and often higher in manufacturing than service sector jobs. The provincial authorities have not wanted such a large migrant population, and if these wage increases push companies to automate processes or relocate production, then this can contribute to the wider strategic objective of taking Guangdong up the value chain. One typical traditional manufacturer of low-end consumer products in Dongguan in early 2011 was dramatically reducing his labour requirements (originally 70 percent of total costs) through automation, buying in machines that could be operated by one person and do the work of twenty workers.[24] Another challenge was that those workers were harder to find, coming from further afield in China than before; in line with the demographic shifts in China they also tended to be older and were more likely to be male than in the two decades after 1992.

For manufacturers, staying in Guangdong meant upgrading through automation, dealing with stricter (enforcement of) environmental regulations, and making efforts to match provincial policy rhetoric by moving up the value chain. Alternatives were simply to shut up shop, as a reasonable number of family-owned Hong Kong-based businesses appear to have done, or to move operations to alternative locations. This meant shifting in one of two directions, either inland to parts of China outside the PRD, or overseas to other Asian economies where wages were lower, and regulation possibly less strict. Those moving inland tended to go to places close to the PRD, though occasionally leaping into Chongqing or up the coast to Shandong, while Vietnam, Bangladesh, Cambodia, and Myanmar were favoured destinations around Asia. In numerous cases the owners were Taiwanese, among the most adept at rapidly shifting production to chase the best deal.

Guangdong outside the PRD: Qingyuan city

A brief foray into Qingyuan, a city in Guangdong just outside the PRD, gives some insights into these dynamics in the early 2010s.[25] Qingyuan is the largest prefecture-level city in Guangdong by administrative area, but its topology is as much rural as urban, with hilly and afforested areas covering well over half of the administrative area. It lies to the north of Guangzhou, and the urban core of Qingyuan is only 70 kilometres away from Guangzhou, not much over an hour's drive by car. Even so, the disparities in economic development with Guangzhou and other parts of the PRD were stark, with notable disparities within Qingyuan between the hilly northern areas of the city and the south, more closely connected to the PRD. A report by the Economist Intelligence Unit on investment locations in China categorises Qingyuan as a 'fourth tier' city within Guandong.[26] At RMB1,010 per month in 2013, up from RMB850 in 2011 and RMB580 in 2008, the minimum wage was no more than two thirds of that in Guangzhou. The demographic profile also reflects this. Whereas (as noted above) Guangdong as a whole is a substantial net recipient of migrant workers from other parts of China, the net inflows of migrants in the southern part of Qingyuan have been small, while workers flow out of the northern areas to the other parts of the province or elsewhere.

In 2011, half of Qingyuan's roughly four million population was living in urban areas. Its GDP was growing rapidly, as was fiscal revenue and investment into the city, mainly from domestic companies such as Anhui Conch (a cement company), and ceramics manufacturers moving from nearby Foshan as that city's economy moved up the value chain. In contrast to Guangdong as a whole, where total trade has consistently been equivalent to over 100 percent of provincial GDP and net exports have been a major driver of provincial GDP growth, the trade to GDP ratio for Qingyuan was less than 30 percent, with net exports in 2011 under USD100 million.

This short detour into Qingyuan highlights two important points about the changing political economy of the PRD and Guangdong. The first is that there is plenty of space still for Guangdong to provide a relatively low-cost location for global production, and places such as Qingyuan offer proximity to familiar export routes in Guangdong, as well as to Hong Kong. Put another way, the incorporation of Guangdong into the global capitalist economy has not yet stretched much beyond the PRD – though this could change. The second point is that the transformations in Guangdong, both before and after the global crisis, have been concentrated in the PRD. Although provincial leaders in Guangdong have attempted to encourage more development in the peripheral areas of the province, there has been limited impact so far.

Transformations across the PRD

By 2013 the rapid economic growth across the PRD had taken the region's GDP (not including Hong Kong and Macao) to USD857 billion, close to that of Indonesia at USD869 billion, with its total trade volumes of USD1,047 billion approaching those of South Korea at USD1,154 billion. The PRD accounted for 80 percent of Guangdong's GDP and over 95 percent of its foreign trade; it accounted for almost 10 percent of China's GDP and about one quarter of the country's trade.[27] In terms of provincial trade volumes, Guangdong remained well ahead of Jiangsu and Zhejiang, though utilized foreign direct investment in Jiangsu in 2011–2013 was USD101 billion compared to Guangdong's USD70 billion.[28]

However, the net result of policy developments within China combined with government and business responses to the global crisis has been less to do with the size and scale of the PRD economy, but more its changing quality and nature. This period has witnessed the transformation and upgrading of the economy across parts of the PRD, partly resulting from the reaction of the Guangdong authorities to the global financial crisis, pushing the closing and relocation of low value-added manufacturing and encouraging higher value-added and service industries. There has been a strong sense that the transition to a new type of economy in the PRD had been realised ahead of other parts of the country. This has even been the case

TABLE 3.3 Guangdong thirteenth five-year programme, key targets for 2020

Provincial GDP	USD1.68 trillion
GDP growth (annual average)	7%
Fixed asset investment (average annual increase)	15%
Retail spend (average annual increase)	9%
Exports (average annual increase)	1%
Air passenger volume	140 million
Railway network	7,000 km
Port handling capacity	2 billion tons
Internet	4G available throughout Guangdong
Metro systems	Guangzhou, Shenzhen, Foshan, Dongguan
New forest planted	3,333 km²
Incomes per capita	Double 2010 levels by 2018
• Urban/rural	• USD7,346/USD2,425
New urban jobs	5.5 million
Hospital beds	6 per 1,000 people

Source: www.newsgd.com/news/2016-03/07/content_143611052.htm

in places such as Dongguan, whose economy was most exposed to trade processing and labour-intensive, low value-added exports, and therefore hit hardest by the crisis, but which has begun to shift into intelligent manufacturing.[29] If the goals set by the provincial government for 2020 are met (see Table 3.3), then the basis for further transformation of the economy looks set to continue.

One means of reducing reliance on exports has been for manufacturers to focus more on domestic consumers, responding to the growth in consumption across China, from major cities to rural areas. The transition has not been without social and economic challenges, though, evidenced for example by numerous incidents of industrial unrest.[30] Other shifts in the structure of the PRD economy have resulted from the growth of clusters in some major industries, in particular the automotive sector and technology companies. The automotive cluster has developed primarily around Guangzhou, building on investment by Japanese companies such as Honda and Toyota,[31] as well as state support for Guangzhou Auto. The supply chain effect has been felt outside the city, including in Qingyuan's efforts to attract investment from the automotive industry.

Shenzhen: China's 'start-up capital'

The technology focus has been in Shenzhen, which has been at the forefront of the transformations in the PRD economy over recent years. Because of the location of the PRD's ports and Shenzhen's historic role in the processing trade, it still accounts for over half of PRD exports, but the more noticeable changes to the economy have come from the growth in domestic consumption and the services sector, and the development of a city that increasingly resembles or surpasses other major Asian cities, and for some has features that make it more attractive than Hong Kong. The property sector has seen rapid increases in prices as a result.

Moreover, Shenzhen is becoming what a number of people have described as China's 'start-up capital' as entrepreneurs from across the country are attracted by its good business and living environment.[32] This combination results to some extent from its location in the temperate climes of the south, but more from the city's recent history as a 'migrant city', one that has attracted people from all over China and in which people have felt able to settle reasonably easily (some other major cities in China are perceived as discriminating against outsiders). Shenzhen attracts talent from across China, with some suggestions that large Chinese companies have been looking to relocate functions and staff to Shenzhen to take advantage of the good business environment and infrastructure. This is reflected in its linguistic culture, based on standard Putonghua without a dominant local dialect (including Cantonese, the main language across most of the rest of Guangdong). This 'migrant city' does not come without problems, however – of Shenzhen's 14 million resident urban population, only three million hold full urban registration (the *hukou* document), though the importance of this may decline under the latest plans to reform the *hukou* system.

From the perspective of technology, Shenzhen is developing advantages in a number of areas, with innovative industries accounting for maybe as much as 40 percent of Shenzhen's economy.[33] Shenzhen is home to a series of leading Chinese enterprises, from telecoms companies Huawei and ZTE and internet giant Tencent to drone-manufacturer DJI and the genome company BJI. Across the region there is an upbeat mood about Shenzhen's entrepreneurial environment, in contrast both to Hong Kong and many other parts of the PRD. Shenzhen spends four percent (or more according to some sources) of its GDP on research and development; both this and its per capita GDP, which reached RMB137,000 in 2013,[34] make it the most prosperous city in China, and put the city in the same bracket as South Korea, substantially above the Guangdong and national averages. Shenzhen also accounts for a disproportionate amount of Guangdong's research and development activity. Across the province this has some particular characteristics: less basic research, but much stronger in application-focused research and development, led by enterprises, and where the vast majority of research and development expenditure is targeted. According to the UK-based innovation non-profit organization NESTA, businesses fund almost 90 percent of research and development in Guangdong, and 45 percent of patent grants in Guangdong are industrial designs (rather than invention patents or utility models). This compares to under half of spend on research and development being done by businesses in Beijing and more like 70 percent in Shanghai; unlike in Shanghai, many of Guangdong's large companies are privately owned and run.[35]

The net result of these changes is a somewhat more domestically-focused PRD economy than in the run up to the crisis, with much of this dynamism coming from servicing and exploiting the Chinese market's scale and domestic consumption, thus diluting the importance of participation in trade processing and production networks. But at the same time, there has been a continued policy emphasis from Guangdong on looking outwards, both building connections with the global economy, and benchmarking itself against other economies as it strives to move up the value chain. These developments reflect the wider drive in China to reach new levels in the global economy.

The pilot free trade zone in Guangdong

This can be seen in approval given in April 2015 for the establishment of the China (Guangdong) pilot free trade zone,[36] one of three new free trade zones approved over a year and a half after the first pilot zone was established in Shanghai. It builds on experimental reforms already piloted in two parts of Guangdong: Qianhai in Shenzhen and Hengqin in Zhuhai. These began under the previous provincial leadership, and the newer free trade zone brings them together with Nansha; the zone itself is therefore made up of three non-contiguous areas: Nansha in Guangzhou at 60 square kilometres, Qianhai and Shekou in Shenzhen at 28 square kilometres, and Hengqin in Zhuhai at 28 square kilometres.

In terms of the macro policy objective, the overall goal is to allow gradual and controlled implementation of measures that are needed to take forward market- and international-oriented reforms. Promoting greater 'openness' is therefore one of the main themes of the policy. Spatially and geographically the priority is to combine greater cooperation with Hong Kong and Macao with playing an important role in national development,[37] all the while promoting global interactions – summed up in the phrase 'depending on Hong Kong and Macao, serving the mainland, facing the world'.[38] Another spatial link mentioned up front is for the zone to act as a 'pivot' for the 21st-century maritime silk road, part of the belt and road initiative. This does not get much treatment in the document other than an injunction for Guangdong, Hong Kong, and Macao to strengthen trade with countries along the silk road. The policy document also suggests that the zone could have spillover effects, which would radiate positively, through the upgrading of processing industries in the PRD, becoming a 'comprehensive services zone' for the PPRD, and establishing a window for outward investment by both enterprises and individuals.

In terms of functional policy measures, the goal is to try to develop an 'internationalised, marketised, rule-of-law-based business environment'. This breaks down into a number of more specific areas, including improving business environment through measures such as the protection of intellectual property rights; liberalising the management of investment (primarily by adopting the 'negative list' approach to investment); further liberalizing trade in services between Guangdong and Hong Kong and Macao; facilitating smoother flows in service sectors by aligning regulations and other measures; promoting trade and developing international logistics and transport services; furthering the cross-border use of the RMB with the goal of making it the main currency for trade, investment and settlement; and other financial services liberalization to match the liberalization in services trade.

The challenge is one of implementation, in particular where other actors – in this case including Hong Kong and Macao – are involved. This relates to more detailed reflection of Hong Kong's role in the economic development of the PRD, which as noted above has been a crucial one since the 1980s. The next section explores how that role is changing and has become more complex and multi-layered over recent years, with shifting hierarchies between Hong Kong and different parts of the PRD. It also highlights the ways in which politics – especially in Hong Kong – features in the relationship between Hong Kong and the PRD, a factor not just in constraining or guiding developments within Hong Kong (which are not the focus here), but as one factor determining the trajectory of this region, and with it the prospects for the Greater Bay Area.

Hong Kong and the PRD[39]

At the same time as reform and opening up stimulated the readjustments in the economic and commercial relationships between Hong Kong and the PRD in the 1980s, a parallel political process was developing, focusing not on Hong Kong and the PRD, but dealing with Hong Kong's status and relationship to (the rest of) China after the envisaged handover to Chinese sovereignty in 1997. The agreements that emerged – the Sino-British Joint Declaration of 1984 and China's Basic Law for Hong Kong of 1990 – were intended to enshrine the commitment of the Chinese government to preserve Hong Kong's separate system after 1997, summarised by the 'one country, two systems' principle, and 'Hong Kong people running Hong Kong' with 'a high degree of autonomy'.[40] The result was that Hong Kong–PRD relations featured a mixture of economic integration with political agreements to preserve the separate nature of Hong Kong's social, economic and administrative system. The interaction of these two features, and the potential tensions between them, provide the basis for many of the developments in Hong Kong–PRD relations since 1997.

After the handover, institutional links between Hong Kong and Guangdong grew gradually, only accelerating somewhat in response to the SARS epidemic in 2003 which not only delivered an economic shock, but also highlighted the need for cross-boundary cooperation to provide for the basic well-being of the population. On the economic side, this led to the Closer Economic Partnership Agreement (CEPA) between the Hong Kong and central governments to give Hong Kong goods, and later services, preferential access to the mainland market, followed by the facilitation of more mainland visitors to Hong Kong. Both helped to support the Hong Kong economy, as noted by then British Foreign Secretary Jack Straw in his comment in 2004 that economic sentiment in Hong Kong 'has improved significantly, not least as a result of continued economic integration with Mainland China.'[41] Indeed, these developments prompted growing talk of economic 'integration' between Hong Kong and the mainland, especially the

PRD, and these ideas were subsequently pushed further by business interests in Hong Kong, with calls to remove obstacles to economic integration and improve the efficiency of cross-boundary flows.[42] Given the separate systems guaranteed in the Basic Law, this trend began to touch some raw nerves in Hong Kong, especially when differences in the two systems were cited as impediments to the emergence of an integrated PRD metropolis.[43]

In response, the approach from the mainland has often been interpreted as seeking to promote integration, though 'integration' has not been the concept deployed in official policy statements. The central government plan for the reform and development of the Pearl River Delta (2008–2020), cited above, talks about enhanced *cooperation* between the PRD and Hong Kong and Macao, based on complementary strengths between the different regions. Deepening economic cooperation (*hezuo*), rather than integration (*ronghe*), between the mainland and Hong Kong is also the theme of the relevant material in the twelfth and thirteenth five-year programmes (for 2011–2015 and 2016–2020 respectively), alongside more general commitments on the part of the central government to support Hong Kong's development. This clearly extends to improving connectivity between Hong Kong and its hinterland, reflecting and responding to growing demand not just from businesses but also from those travelling across the boundary on a regular basis for leisure or family reasons, but without necessarily implying wider integration, with its more political overtones. The latest iteration of the central government's policy statements also includes the concept of the Guangdong–Hong Kong–Macao Greater Bay Area: in his March 2017 annual work report, premier Li Keqiang said that the government would 'draw up a plan for the development of a city cluster in the Guangdong–Hong Kong–Macao Greater Bay Area, give full play to the distinctive strengths of Hong Kong and Macao, and elevate their positions and roles in China's economic development and opening up'.[44]

It is as much social factors as government plans that have driven change in Hong Kong–PRD relations. Recent years have seen a steady growth in cross-boundary marriages,[45] as well as the phenomenon of children living in Shenzhen but travelling through immigration checkpoints on a daily basis to attend school in Hong Kong – it was reported that the number of cross-boundary school buses would increase to 220 in 2016.[46] At the same time as these developments appear to bring communities on both sides closer together, the cementing of a Hong Kong identity that emphasizes difference from that of a mainland identity (itself often homogenized in an overly simplistic fashion) has partly resulted from the diminishing of grassroots cultural and kinship links, which were a big part of Hong Kong's relationship with the PRD from the 1950s onwards.

However, efforts to enhance physical connectivity – especially the construction of the Guangzhou-Shenzhen-Hong Kong high-speed rail link, which comes with a controversially high price tag – have fed back into political sensitivities in Hong Kong, where relations with the mainland have become intensely politicised.[47] Controversy over this project has continued, with critics suggesting that 'one country, two systems' is undermined by the government's proposal to establish both mainland and Hong Kong immigration points at the terminus in Kowloon. This is the result of a combination of fear that the central authorities want to tighten their grip on Hong Kong politics and society and a wider growth in anti-mainland sentiment following the increase in flows of mainland visitors, talent and capital into Hong Kong. While the first of these points is not about the PRD as such, Hong Kong's hinterland is in the forefront of the latter, as seen by the measures taken by the Hong Kong government in 2015 to restrict flows of visitors from Shenzhen into Hong Kong. Another major infrastructure project that has stimulated heated debate in Hong Kong is the 30 kilometre Hong Kong–Macao–Zhuhai bridge. Views on its likely impact vary from those who think it will create new opportunities in linking Hong Kong with Zhuhai and the west bank of the Pearl River (an important part of developing the Guangdong–Hong Kong–Macao Greater Bay Area), to those who see it as a wasteful 'white elephant'.

The political sentiments in Hong Kong do not appear to be fully understood on the mainland side of the boundary, but may have had the effect of curbing the appetite there for greater integration with Hong Kong. Cities in the PRD are inclined to press on with their own strategic development plans, and there is continued infrastructure investment between these cities and beyond, for example in the railway networks up the coast to Xiamen. Rather than pursue complementary integration with Hong Kong airport under which Shenzhen would focus on domestic flights and Hong Kong on international routes, the airport in Shenzhen has been expanding its own international connections. At the same time, though, connectivity is being facilitated through developments such as introducing an option in Guangzhou to check in for flights from Hong Kong airport,[48] while passengers from Hong Kong have for some time been able to check in for flights from Shenzhen airport at a major shopping mall in Hong Kong's Kowloon district and travel by cross-boundary bus directly to Shenzhen airport.

More generally, though, it seems that the Guangdong government still looks to further cooperation with Hong Kong. This was the message sent by Guangdong Party Secretary Hu Chunhua in his March 2016 meeting with the Hong Kong Chief Executive, when he noted that Hong Kong accounted for 20 percent of Guangdong's external trade and was the largest source of investment in the province, over USD20 billion in 2015.[49] Interlocutors in the PRD also frequently say that there is a strong strategic desire on the part of the central and provincial authorities for Hong Kong to retain its advantages, including the rule of law, and to play a role in reforms such as the internationalization of the RMB, though this message has not permeated much of Hong Kong society, where the default assumption seems to be that the central government wants to dispense with Hong Kong's particular status, in spite of continued stated commitments to the 'one country, two systems' principle. Indeed, Hong Kong's comparative advantage benefits mainland counterparts in developing cooperation with or projects in Hong Kong, and many Guangdong companies have set up shop in Hong Kong. There are different emphases within the mainland too, and there may be more enthusiasm for links with Hong Kong on the part of Shenzhen than Guangzhou, due partly to proximity, and also to a sensitivity to potential concerns in Beijing if too powerful a Cantonese economic entity were to develop in the south.[50]

Taken together, these change in Hong Kong–PRD relations have erased the clear hierarchy of the past, when Hong Kong was seen as the more developed, modern city, while the PRD was an unintegrated manufacturing region playing catch up. Now the balance is much more complicated to assess. On a per capita basis, in 2014 Hong Kong's GDP was over 1.7 times that of Shenzhen and twice that of Guangzhou,[51] though Shenzhen's total GDP could surpass Hong Kong's by the end of the decade (see Table 3.4 for data). The continued rapid development trajectory of the PRD means that the gaps are likely to narrow further to 2020. In contrast to the 4 percent of GDP spent on research and development in Shenzhen, the figure for Hong Kong is only 0.73 percent, little of which comes from the private sector. Perhaps reflecting some of these trends, in 2016 the Hong Kong government announced a USD257 million fund to bring venture capital from Shenzhen to Hong Kong to support technology startups.[52]

In terms of livelihood issues, the hierarchies have changed and become more complex too. Living space and some urban planning in Shenzhen are seen by many, both in Hong Kong and Shenzhen, as better than in Hong Kong. On the other hand, Hong Kong's public education and health facilities are favoured, and for expatriate managers in particular there is still a preference for living in Hong Kong over Shenzhen. Hong Kong is also perceived as having much to offer in terms of institutional experience, human resources, and international outlook and

TABLE 3.4 Comparison of economic data (2014)

	Shenzhen	*Guangzhou*	*Hong Kong*
Land area (km²)	2,050	7,435	1,108
Population (mn)	10.79	13.08	7.24
GDP (RMB bn)	1,600	1,671	1,882
GDP per cap (RMB)	149,495	128,478	260,000
Primary industry/GDP	0.0%	1.3%	N/A
Secondary industry/GDP	42.6%	33.5%	N/A
Tertiary industry/GDP	57.4%	65.2%	c.93%★
Exports (USD bn)	284.4	72.7	470.9★★
Imports (USD bn)	203.3	57.9	540.9
Foreign direct investment (USD bn)	5.8	5.1	103

★ From www.gov.hk/en/about/abouthk/factsheets/docs/service_economy.pdf.

★★ Of this USD463.8 billion (98 percent) was re-exports. For discussion of the complexities in analyzing this data, see notes in InvestHK, *The Greater Pearl River Delta* (Hong Kong: Hong Kong SAR Government, 2014), p. 18.

Source: Hong Kong Trade Development Council, data taken from various research reports available at www.hktdc.com/Research. Figures for Hong Kong in US dollars converted to RMB at USD1=RMB6.5.

contacts, and these are areas where Shenzhen and Guangzhou are seeking to engage and catch up, including through pilot projects such as the Qianhai Shenzhen-Hong Kong Modern Service Industry Cooperation Zone, where half of registered companies are in financial services, though interlocutors from the Shenzhen side note that – with a few exceptions such as HSBC – this has become essentially a Shenzhen project due to a passive response from Hong Kong.

Another area where there is clear agreement that Hong Kong retains an advantage is in the rule of law, increasingly cited by business people in particular as its single most important point of comparative advantage (the days when Hong Kong's main advantage was seen to be its people seem to have faded). The fact that this is based on a different legal system and tradition from the mainland has the potential to complicate cross-boundary cooperation and commercial interconnectedness, including through the Closer Economic Partnership Agreement between Hong Kong and the mainland. It means that full integration would be challenging to achieve, though rather than being an 'obstacle', interlocutors in the mainland suggested it could offer opportunities to make use of the comparative advantage of Hong Kong.

Conclusion

The PRD is a region that has been redefined and formed in the context of China's integration into the global economy, which took place at a time when the country was entering a new phase in its policy approaches and development. This has seen the region rapidly transform from a relatively less developed part of the country to one of its most prosperous, off the back of relationships with Hong Kong as the then British territory shifted away from actual manufacturing activity towards its management, and then to business and financial services. This trajectory is a good example of 'a conjunctural accident of world-historical significance', based not solely or even primarily on policy or business decisions made within China, but the ways in which multiple actors outside and in this part of China responded to global change. Over the last decade it has again been global developments – the financial and economic crisis and its ongoing aftermath – which have spurred a new phase of transformation across much of the region, and the imagined future reshaping of the region into a Greater Bay Area. This time the Chinese economic context is very different from that in the 1980s: rather than needing to make a 'desperate move to ensure China's economic security', China's economy is well on its way to becoming the world's largest in aggregate terms.[53]

This transformation indicates some of the important shifts in China's relationships with the global economy, and therefore – given China's scale and levels of global integration – in the trajectories of globalization itself. The rise of the Chinese consumer has given more sustainability to domestic drivers of the economy, but also demonstrated that China's integration into global capitalism is about more than playing a role at the middle (or lower) levels of global production networks.[54] Shenzhen and Guangzhou are at the forefront of this. Further, a number of companies, in Shenzhen in particular (along with some from other parts of the country), have become among the first to challenge Euro-American-Japanese dominance of the 'commanding heights' of the global economy.[55] China's outward investment is often thought of in terms of debt-financed acquisitions, and these are fast becoming a central factor in change in the global economy, but it is businesses such as Huawei that appear to be charting a more sustainable and organic path to playing a leading role in their sectors, though it remains to be seen whether this will be at the technological frontier or as fast followers. And outsourcing by traditional manufacturers in Guangdong to locations in inland China or elsewhere in Asia further reinforces the region's role in ongoing shifts in value chains and production networks.

Looking again at the idea of the Greater Bay Area – or the PRD mega-city first discussed in the 1990s – shows the importance of the global linkages in all of these developments. Growing connectivity and integration has been partially realised within the region as it has developed into a multi-core networked spatial entity, but as discussed above, the political dynamics around Hong Kong's relationship with the rest of China and their implications for the Hong Kong–Guangdong boundary have constrained the organic integration of Hong Kong (and Macao). Nonetheless, connectivity has clearly been strengthened, in social relations, through the rapid growth in visitors to Hong Kong from mainland China (many from Guangdong), and in the continuing division of labour within the global economy. Hong Kong's continued advantages in international business and financial services, which are particularly brought to bear in supporting outward investment by Chinese enterprises, mean that Hong Kong's growing connectivity to mainland China is only partly territorial, through its links to its hinterland in Guangdong, but remains most significant in ways that take advantage of its different system, operating under the 'one country two systems' formula. For those who follow Hong Kong politics, this means that Hong Kong is clearly not becoming 'just like any other city in China', contrary to claims made by many in Hong Kong, and echoed in commentary further afield.

All of this reflects an important theme in post-1978 regional policy in China, that of the central role given to utilizing comparative advantage between different regions. The particular features of the PRD and Hong Kong's continued differentiated role and separate system all create the space for regions to contribute to a diverse and more resilient relationship between China and the global economy. Finally, looking at this corner of China suggests that we are in a 'post-PRD' era where alternative ways of looking at and dealing with the region are more useful in thinking about the global interactions of this area. Rather than focus on the region as a 'surface', we have seen a shift to multi-level differentiated networks within the region, but also extending well beyond, perhaps even along a revived maritime silk road, if that comes to fruition. We return to that question in discussion of the belt and road initiative, but first head inland to the municipality of Chongqing.

Notes

1 Justin Yifu Lin and Peilin Liu. 'Development strategies and regional income disparities in China'. In Guanghua Wan, ed. *Inequality and growth in modern China*. Oxford, UK: Oxford University Press, 2008, pp. 56–78, p. 56.
2 Anita Chan, Richard Madsen, and Jonathan Unger. *Chen village: revolution to globalization*. Berkeley, CA: University of California Press, 2009.
3 Tian Xiaowen. 'Deng Xiaoping's *Nanxun*: impact on China's regional development'. In John Wong and Zheng Yongnian, eds. *The Nanxun legacy and China's development in the post-Deng era*. Singapore: Singapore University and Press and World Scientific, 2001, pp. 75–94. In 1988 the *Far Eastern Economic Review* had already referred to Guangdong as 'Asia's fifth dragon'; see: Dali Yang, 'Patterns of China's regional development strategy'. *The China Quarterly*, 122, 1990, p. 251, footnote 56.
4 Carolyn Cartier. 'Uneven development and the time/space economy'. In Wanning Sun and Yingjie Guo, eds. *Unequal China: the political economy and cultural politics of inequality*. New York and Oxford, UK: Routledge, 2013, p. 87.
5 This English translation is cited from Bill Overholt in: Chen Xiangming. *As borders bend: transnational spaces on the Pacific Rim*. Boulder, CO: Rowman and Littlefield, 2005, p. 64.
6 Alvin Y. So. ' "One country, two systems" and Hong Kong–China national integration: a crisis-transformation perspective'. *Journal of Contemporary Asia*, 41: 1, 2011, p. 105.
7 From 1979–1990, 45.5 percent of utilized FDI into China went to Guangdong; Michael Enright, *Developing China: the remarkable impact of foreign direct investment*. London and New York: Routledge, 2016, p. 26.
8 Qu Baozhi. 'Mainland China–Hong Kong economic relations'. In Joseph Cheng, ed. *The Hong Kong special administrative region in its first decade*. City University of Hong Kong Press, 2007, p. 409.
9 Dali Yang. 'China adjusts to the world economy: the political economy of China's coastal development strategy'. *Pacific Affairs*, 64: 1, 1991, p. 44.
10 See Chapter 1 of the present book; also Milivoje Panić, *Globalization: a threat to international cooperation and peace?* Basingstoke, UK and New York: Palgrave Macmillan, 2011.
11 Kjeld Allan Larsen. *Regional policy of China, 1949–85*. Manila: Journal of Contemporary Asia Publishers, 1992, pp. 191–193; Chen, *As borders bend* (see note 5), p. 75.
12 Wang Shaoguang. 'The rise of the regions: fiscal reform and the decline of central state capacity in China'. In Andrew G. Walder, ed. *The waning of the Communist State: economic origins of political decline in China and Hungary*. Berkeley, CA: University of California Press, 1995, pp. 87–113.
13 Gerald Segal. 'The Muddle Kingdom? China's changing shape'. *Foreign Affairs*, 73: 3, 1994, p. 49.
14 See discussion in Chapter 2 of the present book.
15 Qu, 'Mainland China–Hong Kong economic relations' (see note 8), p. 409.
16 Chen, *As borders bend* (see note 5), pp. 72–74.
17 This is as a 'network' not a 'surface' in the language used in Chapter 1.
18 For this paragraph and quotations, see: Manuel Castells. *The rise of the network society*. Oxford, UK and Malden, MA: Blackwell, 2010, pp. 434–440.
19 Both Shanghai and Beijing recorded higher trade/GDP ratios, but the concentration of major national state-owned enterprises in both cities distorts a direct comparison. Author's calculations from data in China Statistics Press, *Zhongguo diqu jingji jiance baogao* (*Monitoring report on China's regional economy*). Beijing: China Statistics Press, 2010.
20 Presentation attended by the author in 2009.
21 Dali Yang. *Beyond Beijing: liberalization and the regions in China*. London and New York: Routledge, 1997.
22 The term for crisis in Chinese (危机 *weiji*) is composed of two characters representing 'danger' and 'opportunity'.
23 These ideas somewhat foreshadow the change made in the 'principal contradiction' facing the Party at its 19th National Congress in October 2017.

24 Author's visit.
25 This material is based on two visits the author paid to Qingyuan during this period.
26 Economist Intelligence Unit. *China hand: the complete guide to doing business in China 2014*, available through www.eiu.com, 2014, p. 21. Minimum wage data for 2011 and 2008 are from the author's visits to Qingyuan.
27 HSBC, 'Asia – Uniquely positioned to capture growth in the Asian century', investor update, 9 June 2015, www.hsbc.com/~/media/hsbc-com/investorrelationsassets/ investor-update-2015/asia-china-and-the-pearl-river-delta, slide 13.
28 Trade data from China Statistics Press, *China statistical yearbook 2015*; data on FDI from Enright, *Developing China*, p. 26.
29 Guangdong News. 'Dongguan on front lines of robot revolution', 2 March 2016, www.newsgd.com/news/2016-03/02/content_143342870.htm. This paragraph also draws on the author's discussions in early 2016 with people engaged in policy making and business in the region.
30 See data on strikes produced by non-governmental organization China Labour Bulletin, available at http://maps.clb.org.hk/strikes/en.
31 Economist Intelligence Unit, *China hand* (see note 26), pp. 27–28.
32 Author's research, mainly conducted in 2016.
33 Week in China. 'Belt and road: China's grand gambit', 2016, www.weekinchina.com/ wp-content/uploads/2016/11/WiCFocus13_4Nov2016.pdf.
34 Economist Intelligence Unit, *China hand* (see note 26), p. 29.
35 NESTA. 'China's absorptive state: innovation and research in China', October 2013, www.nesta.org.uk/publications/chinas-absorptive-state-innovation-and-research-china, pp. 39–43.
36 State Council. 'Guowuyuan guanyu yinfa Zhongguo (Guangdong) ziyou maoyi shixianqu zongti fang'an de tongzhi' (Notification from the State Council about the approval of the overall programme for the China (Guangdong) pilot free trade zone), April 2015, www.gov.cn/zhengce/content/2015-04/20/content_9623.htm.
37 Though Macao is mentioned here and in similar sources, the main focus is on Hong Kong.
38 依託港澳 服務內地 面向世界 (*yituo gang'ao, fuwu neidi, mianxiang shijie*). All translations are the author's.
39 This section is partly based on interviews in Hong Kong, Guangzhou and Shenzhen carried out in early 2016 as part of a research project on behalf of the Asia programme at Chatham House, funded by the Foreign and Commonwealth Office. I am grateful to both organizations for allowing me to draw on material from that research in this chapter.
40 For a recent discussion of the implementation of these ideas see: Tim Summers. 'Hong Kong: is the handover deal unravelling?' In Kerry Brown, ed. *The critical transition: China's priorities for 2021*. London: The Royal Institute of International Affairs, 2017, pp. 11–15, www.chathamhouse.org/publication/critical-transition-chinas-priorities-2021.
41 Foreign and Commonwealth Office. *The six-monthly report on Hong Kong, 1 July to 31 December 2003*, deposited in Parliament by the Secretary of State for Foreign and Commonwealth Affairs, p. 1.
42 Bauhinia Foundation Research Centre. 'Creating a world-class Pearl River Delta metropolis: accelerating economic integration between Guangdong and Hong Kong', August 2008, www.bauhinia.org/assets/pdf/research/20081028/PRD%20full%20report-en.pdf.
43 Chen, *As borders bend* (see note 5), p. 77.
44 Li Keqiang, 'Report on the Work of the Government', 16 March 2017, http://news. xinhuanet.com/english/china/2017-03/16/c_136134017.htm.
45 Bauhinia Foundation. '"Hun-li" yuanlai buyi, zhonggang huren you fayi?' ('"Marriage and divorce" is not easy, is there a legal basis for mutual recognition between China and Hong Kong?'), 6 August 2016, www.bauhinia.org/index.php/zh-HK/analyses/476.

46 Guangdong News. 'HK to bring in private venture capital from SZ', 2 March 2016, www.newsgd.com/news/2016-03/02/content_143324774.htm.
47 The Economist. 'Over troubled water: cross-border transport links are overshadowed by political fears', 13 February 2016, www.economist.com/news/china/21692935-cross-border-transport-links-are-overshadowed-political-fears-over-troubled-water.
48 Guangdong News. 'Check-in service for Hong Kong International Airport to open in downtown Guangzhou', 3 March 2016, www.newsgd.com/news/2016-03/03/content_143409750.htm.
49 Hong Kong Government. 'CE meets Guangdong officials', 5 March 2016, www.news.gov.hk/en/categories/admin/html/2016/03/20160305_193711.shtml.
50 See Segal,'The Muddle Kingdom?' (see note 13), pp. 43–58, for some earlier discussion of these issues.
51 Author's calculations based on data from city profiles provided by Hong Kong Trade Development Research on www.hktdc.com, with data for Hong Kong given in US dollars converted to RMB at USD1 = RMB6.5.
52 Guangdong News. 'HK to bring in private venture capital from SZ'.
53 Both quotes are taken from: David Harvey. *A brief history of neoliberalism*. Oxford, UK: Oxford University Press, 2005, p. 120.
54 LeAnne Yu. *Consumption in China: how China's new consumer ideology is shaping the nation*. Cambridge, UK: Polity, 2014.
55 Peter Nolan. *Is China buying the world?* Cambridge, UK and Malden, MA: Polity, 2012.

4
CHONGQING
Globalization moves inland

The municipality of Chongqing, located inland in central-western China, is a clear example of a part of China that has sought to develop its economy in response to the country's engagement with an intensified globalization.[1] Chongqing's transformations since the 1990s have been huge, and its engagement with globalization began with the development of an internationally-oriented economic and industrial strategy. This has evolved over time, and the city's most recent global positioning feeds directly into aspirations under the belt and road initiative to build 'silk roads' of connectivity across Eurasia, as well as into relationships with the nearby city of Chengdu and along the Yangzi river. Chongqing came to global prominence in 2012 due to scandals around its controversial party secretary, Bo Xilai, who was seen by some as seeking a new socialist path for China's development, though his promotion of Chongqing's engagement with a capitalist globalization has had a more lasting impact on the city. In sum, Chongqing's story is one of a clear effort to engage with globalization in a way that benefits the municipality.

Climbing the political ladder

Chongqing is located on the upper reaches of the Yangzi river, more than 1,400 kilometres inland from Shanghai. It nonetheless has a history of global interactions stretching back at least to its designation as a treaty port open to foreign traders in the late Qing dynasty, though the British consuls who were posted there struggled to make the tough journey up the river, the only way of accessing the city.[2] Chongqing came to international prominence again in the late 1930s when Chiang Kaishek moved his government inland following the Japanese invasion of eastern China. Its spell as China's wartime capital attracted foreigners such as the American Graham Peck, who wrote movingly of the city in *Two Kinds of Time*.[3]

Following the war and the Communist revolution of 1949, however, Chongqing's links with developments outside China's borders faded. After initially maintaining its separate status reporting directly to the central government, the old city was incorporated into Sichuan province again after the abolition of Greater Administrative Regions in 1954, though it retained a degree of autonomy from the province in economic planning for parts of the next three decades. The security attributed to its remoteness and hilly terrain made Chongqing a key node in the Third Front strategy launched in the mid-1960s, bringing to the mountainous environs of the city a range of heavy defence and industrial bases whose legacies are still present in its economy today; during the Third Front period, Chongqing accounted for 200 of the 700 projects relocated inland from the coastal regions. But for several decades Chongqing's overall status was as the second city in Sichuan province, and as result of this – and the inefficient nature of Third Front investment – its overall economic development lagged behind that of the provincial capital Chengdu.[4]

Urban reforms piloted by then Sichuan party secretary Zhao Ziyang were launched in Chongqing in 1978, and in 1983 the city was placed under the direct economic control of the central government as the first city with separate planning status (计划单列市 *jihua danlie shi*), though remaining within Sichuan's political and administrative jurisdiction.[5] As a result of this change in status, Chongqing was effectively at provincial level for economic planning, accountable directly to the State Planning Commission, with the intention that it would be 'China's most important experimental zone for the restructuring of the economic management system'.[6] The flexibility enabled its government more easily to redevelop economic links with adjacent areas of neighbouring provinces (Yunnan, Guizhou, southern Shaanxi, southern Gansu, western Hubei, and north-western Hunan), and thus play more of a role as a regional economic centre. The city area was extended at the same time, from 9,848 to 22,341 square kilometres, giving it a population approaching 14 million, the largest of any administrative city in China, though only 2.7 million lived in the city proper. Chongqing has a strong self-identity as a city of hard-working people, as well as an open place, based on a history of inward migration from many other parts of the country (Table 4.1 gives some current economic data).

It was the planning for the eventual construction of the Three Gorges Dam in the 1990s that brought the city firmly back to the attention of the central Party-state. The project required the resettlement of around one million residents of the area downstream from Chongqing on the Yangzi river, in what were then the far eastern counties of Sichuan province (the dam itself is in the neighbouring province of Hubei). To facilitate this, as well as to explore more effective administration of the province's huge population and more effective development, in 1997 a new municipality of Chongqing was created out of four prefecture-level regions of Sichuan, to report directly to the central government (in Chinese the shorthand for this type of municipality is 直辖市 *zhixiashi*). The population of what was in effect a province with enhanced political status – the same administrative status as

TABLE 4.1 Chongqing economic data

Year	1997	2014
Resident population (million)	28.7	29.9
Urban population (proportion)	30%	60%
Nominal GDP (billion RMB)	151	1,426
Nominal GDP/capita (RMB)	5,253	47,850
Total social consumption (billion RMB)	56.8	571.1
Utilized FDI (million USD)	385	4,233
Total foreign trade (million USD)	1,678	95,450

Source: Data from Fung Business Intelligence, 'Chongqing: Changjiang shangyou de zhongxin chengshi' ('Chongqing: a central city on the upper reaches of the Yangzi'), 2016, www.fbicgroup.com/sites/default/files/Clusters_Series_special_Chongqing.pdf.

Beijing, Shanghai, and Tianjin – was then 30 million, with an urbanization ratio of 30 percent, and its area 82,300 square kilometres, slightly larger than that of Scotland. With municipality status came growing investment from domestic sources, and after the announcement of the Develop the West policy in 1999, the central government and the city's leadership saw the opportunity to turn Chongqing into a 'gateway to western China' and a leading city in central-western China. Its location in 'underdeveloped' western China acted as a tool for local policy makers to seek preferential policies from the central government, and more recently Chongqing officials have been keen to note that the city is at the top of many rankings among cities in *western* China, highlighting that in many ways, Chongqing's development is an example of the trajectory of western China from 2000 onwards. By 2014 the urbanization ratio was 60 percent of population, compared to 30 percent in 1997, though of the 30 million population, at least four million were working as migrants away from their homes but still within Chongqing, and an additional four million registered Chongqing *hukou* holders were living as migrants in other parts of China.[7]

Developing an international economic and industrial strategy

From the 1990s there were signs of growing international business and government interest in Chongqing. Japanese motorcycle and automotive companies had been attracted by Chongqing's growing cluster of motorbike and truck manufacturers, while a joint venture by Ford motor company, and investments by energy major BP in a chemicals joint venture and by UK-based brewer Scottish and Newcastle in the Chongqing Beer company added to the foreign direct investment in the city. In the 1990s the Japanese and Canadian governments opened offices in the city, followed in March 2000 by the establishment of a British consulate-general.

However, for a long time Chongqing remained peripheral to China's global economic interactions. Foreign direct investment was seen locally as a way of modernizing and developing the city, but it was limited. The big challenge was similar to that faced by other parts of western and inland China: foreign investment was either driven by prospects of selling to local markets, which were more attractive in richer coastal cities, or by manufacturing for export. For the latter, Chongqing's distance from coastal ports made it unattractive, even with substantially lower land and labour costs. Neither did Chongqing enjoy a location that could easily support land-based trade with neighbouring countries in Asia – or in some cases even with neighbouring provinces. These questions of logistics, transport networks, and connectivity were near the top of the agenda of the city's policy makers as they addressed Chongqing's economic strategy in the years after Chongqing became a provincial-level municipality in 1997.

In the early 2000s, the envisaged solution in the city to Chongqing's logistical challenges was the Yangzi river itself. Chongqing policy makers asserted that, after the completion of the Three Gorges Dam around 2009, ocean-going container ships (up to 5,000 deadweight tons) would be able to travel up the Yangzi and dock at Chongqing, opening up the city as a lower-cost location for the export-oriented industries and trade processing that had so far clustered on the coast. These ideas were shared by other cities along the Yangzi, and since the dam's completion there has been some sign of this potential being realised, with vessels travelling to Chongqing utilizing massive ship locks on the dam itself. Shipping routes now account for as much as three quarters of Chongqing's exports,[8] though the Yangzi's capacity remains constrained. Over the last decade, therefore, the Chongqing government has concentrated more of its efforts on developing logistics linkages by road, rail and air (discussed further below).

The other element in furthering international trade and investment has been industrial development, which would facilitate the manufacture and export of products from Chongqing. The city's traditional industrial base – much of it a legacy of the Third Front – consisted of heavy and military industries. Based on that legacy, under a programme to convert military to civilian industries from the 1990s, Chongqing saw the emergence of factories to produce industrial equipment, in particular around the automotive and motorcycle industries, where it ranks number three and number one respectively in China by scale.[9] The most prominent automotive enterprise has been ChangAn, owned for a long time by the Southern Industries Group, which has developed its products and know-how through joint ventures with Ford, Mazda, and Suzuki. Chongqing has also long been China's leading centre for the manufacture of motorcycles, accounting today for one third of China's manufacturing and half of its exports in the sector. The city's experience here reflects a wider story of the relationship between Chinese industrial development and the global economy. Motorcycle manufacturers in Chongqing exemplify the ways that Chinese companies have innovated through modularizing production, taking the 'nonmodular, highly integrated system of new product development' that was the characteristic of the Japanese companies with which they entered joint ventures, and 'covert[ing] it into a highly modularized, highly deverticalized mode of production'.[10] Companies in Chongqing such as Lifan,

Jialing, Jianshe, Zongshen, and others were thereby able to lower production costs substantially, and produce motorcycles that were competitive not just in China but in many southeast Asian countries as well.

Nonetheless, many of these traditional industrial sectors were not seen locally as the best avenue for Chongqing to promote exports beyond developing markets such as those in southeast Asia. Neither did Huang Qifan, Chongqing's ambitious executive vice mayor from 2001 and mayor from 2010 to 2017, want to base industrial strategy on low value added light industry, such as that which had fuelled early phases of growth in Zhejiang or Guangdong.

In line with the national macro-economic policy injunction to move up the value chain and develop high-tech industries, Huang alighted on the growing production of consumer electronics in China, in particular netbooks, laptops, printers, and mobile phones. Wooing both HP and the leading Taiwanese contract manufacturer Foxconn (operating in China as Hon Hai Precision Technologies), the Chongqing government negotiated a USD 3 billion investment in 2010 under which HP netbooks would be assembled in Chongqing for export to European markets, initially by air and subsequently over land.[11] Lower labour costs were a big incentive for the manufacturers, but there is more to be said about the features of this arrangement, which give insights into Chongqing's fast-changing engagement with the global economy.

As well as developing a new consumer electronics 'pillar industry' in Chongqing, Huang's strategy looked to bring processing trade to Chongqing, making use of relatively good and low-cost access to land and labour in the city. For Huang, however, there were to be some important innovations in Chongqing's engagement with processing trade. The model for China's processing trade from the 1980s had been to 'import raw and semi-finished materials from the international market and then export the finished products'.[12] Instead of this placing 'both ends outside (*liangtou zaiwai*)' of China, Huang's plan was to place 'one end outside and the other end inside' (*yitou zaiwai, yitou zainei*). This meant bringing as much of the supply chain as possible inside China, preferably in Chongqing, thus benefiting the local economy and avoiding what Huang described as the waste and inefficiency of the 'big imports and big exports' model of an extended supply chain.[13] Although there were grumblings in Chongqing at the time that the value added in the city was minimal, Huang and other policy makers were keen to bring as many of the more profitable types of production activity as they could, influenced by models such as the 'smile curve' developed by Stan Shih of Acer computers, which shows the relative value added and profitability at different stages in the production process.[14] Particularly important in this is engaging in the research and development parts of the process; although this is another area the Chongqing government has sought to develop, the challenges are greater.

Another development of the original 'both ends outside' concept appeared in remarks made by Huang back in 2009. He talked about a 'copy-cat' (*shanzhai*) model of trade processing, which would see a full shift from 'both ends outside' to 'both ends inside'. In other words, not only would the supply chain be brought inside China (and preferably Chongqing), but the final consumption of products would also take place within China's borders. What Huang called this 'real model' of 'Chinese domestic demand trade processing' (*Zhongguo neixuxing jiagong maoyi*) would be driven by future growth in the domestic consumer economy,[15] though as later developments attest, export routes beyond China's borders remained an important priority for the Chongqing government. And in Huang's subsequent rhetoric, it was the 'one end outside, one end inside' concept that dominated.[16] A key part of the deal to attract HP to Chongqing was to develop new logistics routes for export of product to Europe.

Huang also claimed a 'social' dividend, by arguing that this approach to trade processing would be a better for workers than the model that relied on migrant labour moving to the coast (this was shortly after a number of worker suicides at Foxconn plants elsewhere in China), and would facilitate the much talked-about reform of the household registration (*hukou*) system being trialled in Chongqing at the same time, under which rural residents could more easily be given full urban status, and with it a job in a factory assembling consumer electronics (see below). Huang sought a further innovation, by bringing trade settlement to Chongqing as an 'inland offshore financial settlement centre'.[17] In another set of comments, Huang talked about 'three ends' involved in trade processing rather than two, with the third being trade settlement – his strategy would therefore bring 'two ends' inside China, while the third remained outside.[18]

The reality is that both product development and manufacture have become transborder processes and so neither 'end' of the production and consumption process takes place fully inside or outside China, though some in Chongqing suggest that as much as 80 percent of the parts for consumer electronics are now produced locally. Nonetheless, these innovative approaches to trade processing under Huang demonstrate changes in Chongqing's relationship to globalization, bringing Chongqing in a new way into the structures of globalized production similar to those that had facilitated China's earlier integration into the global economy, and in turn driving the development of new logistics routes in order that costs and ease of transportation would not be an impediment to these developments. The broader changes in the Chinese economy that have enabled more of the supply chain and a greater proportion of final consumption to be brought inside China's borders have further implications for the relationship between China and globalization, and maybe even for the nature of globalization itself.[19]

As far as Chongqing is concerned, Huang's policies appear to have succeeded in turning consumer electronics assembly from nothing into a local 'pillar industry' accounting for 19 percent of industrial output in 2015, and in placing Chongqing on the global map in the sector, along with Chengdu, where Foxconn have also made substantial investments. In 2014 and 2015, 40 percent of notebooks globally were assembled in Chongqing, with five major global brands and six Taiwanese contract manufacturers present in the city, along with 800 enterprises involved in the supply chain. Processing trade was also the main driver of Chongqing's foreign trade. In 2015, 76 percent of exports and 82 percent of imports were in mechanical and electronic equipment, and Chongqing's trade was the largest among provinces in western China.[20]

Further industrial policy focused on the desire to move up the value chain – to raise Chongqing's position in the 'international division of labour' – and develop into new industries, reflecting national priorities set out in the State Council's Made in China 2025 document and policy focus on internet-related industrial development. This goal was affirmed by Party general secretary Xi Jinping when he visited Chongqing in January 2016. Sectors prioritized by the Chongqing government include robotics and new-energy automotives. The government has also sought to develop trade in services, with a new plan announced in August 2015.[21] As across China, Chongqing's trade in services had lagged trade in goods (the thirteenth five-year programme contains a national target for trade in services to reach 16 percent of total trade by volume by 2020). In 2010, Chongqing's trade in services was USD3.5 billion, rising to USD13.1 billion in 2014 and USD2.85 billion in the first three months of 2016, up 18.8 percent on the same period in 2015. The government plan projected a total of USD50 billion by 2020. As at the national level, Chongqing's trade in services was in deficit, with exports of USD5.8 billion and imports of USD7.3 billion in 2013.[22]

Chongqing and national policy experiments

One of the wider policy themes to emerge from this account of the development of Chongqing's industrial policy was the willingness and ability of the municipal government to experiment and innovate to address social and economic challenges. One of these areas was in the reform of the household registration (*hukou*) system, under which households are registered in a particular locality, and categorized as either urban or rural, with implications for access to public services such as education and healthcare. Chongqing's *hukou* pilot reforms, and another pilot in rural land ownership – developed after comments in March 2007 by Party General Secretary Hu Jintao during the annual meetings of the people's congresses in Beijing and subsequently encapsulated in a State Council document of 2009, 'Guiding opinions of the State Council regarding further comprehensive urban–rural reform and development of Chongqing municipality' – supported Chongqing (along with Chengdu) becoming a pilot zone for comprehensive reforms in urban–rural coordinated development. As noted above, these reforms were used by the Chongqing government to facilitate the development of new assembly and production by providing a new source of urban, industrial workers.

A subsequent major policy announced for Chongqing was the establishment on 18 June 2010 of the Liangjiang New Area, the third 'new area' to be approved nationally after Shanghai Pudong and Tianjin Binhai.[23] This built on the earlier establishment of bonded customs zones in Chongqing (at Xiyong and Cuntan) to facilitate exports by air, rail, and river. The five main aims for the Liangjiang New Area were to act as a pilot area for coordinated urban and rural reform, an 'important base' for advanced manufacturing industry and modern service industry, a financial and innovation centre on the upper reaches of the Yangzi river, an 'important door for opening up in inland China', and a window to demonstrate the 'scientific concept of development' that was promoted by then Party general secretary Hu Jintao.[24] Its area of 1,200 square kilometres covered almost 40 percent of Chongqing's downtown area and part of the motivation for this administrative innovation was to streamline bureaucracy and enhance integration across district boundaries. Chongqing was therefore becoming a key location for national policy pilots, as well as for initiatives developed at the local level, such as the move into trade processing and logistics connectivity across Eurasia, as well as plans by Huang to develop Chongqing into a cloud computing centre from 2010 onwards.

More recent policy experiments have covered SOE reform, where in both 2014 and 2015 significant equity in SOEs was sold under the guidance of the Chongqing State-owned Assets Supervision and Administration Commission.[25] This built on a target announced in 2014 that two thirds of the city's SOEs should be under mixed state and non-state ownership. Another area is the implementation of the structural supply-side reform, which was first pushed by the national leadership in December 2015.[26] When a further seven free trade zones across China were announced in September 2016, Chongqing was one of the cities included.

A further development with an international element was the announcement in November 2015 of the China–Singapore (Chongqing) Demonstration Initiative on Strategic Connectivity. This is the third Singapore government-led project in China, following earlier ones in Suzhou (1994) and Tianjin (2008), and targets connectivity across finance, telecommunications, aviation and logistics; the central government's focus on it was reflected in a reference in the thirteenth five-year programme.[27] Meanwhile Chongqing companies themselves were looking to 'go global' by investing overseas and developing their international trade. Major state-owned companies such as Chongqing Machinery have acquired companies overseas, while non-state enterprises such as automotive and motorcycle group Lifan have exported products and set up assembly plants across developing economies from Vietnam to Uruguay.

Global positioning: connections across Eurasia

A key priority for the Chongqing government to support its wider economic and industrial policy has been investment in infrastructure and logistics, which has played a central role in opening up Chongqing to greater engagement with the global economy and thereby globalizing the city's industrial and economic development. Funds have been directed to build road and rail linkages to neighbouring provinces, open up potential routes to China's borders and beyond to external markets, develop logistics capacity such as an intermodal port terminal at Guoyuan, and invest in cargo and passenger capacity at Chongqing's airport. Passenger capacity in 2016 was 45 million journeys per year, compared to three million in 2000, since when three new terminals have been constructed.

The global scale of these plans was ambitious, and when it comes to international connections there have been three geographical or spatial strands to Chongqing's positioning. The first is the most significant and has attracted attention not just in China, but beyond after its inclusion in the belt and road initiative. This is the development of freight train services from Chongqing northwest to Xinjiang and then into Kazakhstan at Alashankou, China's busiest land port, into central Asia and on to Europe; hence it is known as the Chongqing (Yu) – Xinjiang (Xin) – Europe (Ou) or Yuxinou route. It passes through six countries across a total distance of 11,179 kilometres. This logistics connectivity was partly stimulated by the agreement for HP products to be assembled in Chongqing, but had a longer-term strategic objective of facilitating other exports to Europe by reducing by half the time taken from some 30 days by ocean transport (which includes time to reach coastal ports from Chongqing).[28]

The first pilot train travelled from Chongqing to Alashankou in five days in October 2010, and following the signing the following month of a customs agreement between China, Russia, and Kazakhstan to facilitate single customs inspections, the first cross-border trains were successfully launched in early 2011, first to Russia, and subsequently to Duisburg, Germany. In 2011, 17 trains departed Chongqing, increasing to 41 in 2012, and a total of 233 freight trains in 2014 and 257 in 2015, a figure projected to reach 700 by 2020. According to the Chongqing government, a total of 614 trains had travelled by the end of May 2016; in the first five months of 2016 there were 84 outbound and 40 return trips. By the end of 2012, a total of over USD 1 billion of goods had been exported by train, including some four million HP notebooks; the total value of goods taken along the route reached USD 6.8 billion at end of 2014.[29]

Unsurprisingly, the benefits of the route have been promoted by the Chongqing government, though these are not without challenges. Countries along the route use different size railway gauges, meaning that trains have to stop and be re-wheeled en route. Changing climatic conditions on the route mean that the temperature of goods needs carefully regulating. In the early days, there were no goods imported along the route, reducing efficiency and increasing costs, though in March 2013 the import of automotive parts for Ford began, and in 2014, the central government authorized Chongqing as a destination for car imports and a first consignment of 80 cars was imported using the train. The costs and profitability of the route – which is operated by a specially-established company – remain unclear. Although the government's figures stress the speed when compared to ocean transport, some logistics consultants say that the costs of the train route mean that it is more sensible to think of it as a substitute for air cargo not sea cargo. Nonetheless, the public response to the train route from global logistics companies has been positive, with comments on the potential for 'tremendous growth in rail freight between Asia and Europe'.[30]

Subsequently, the route received affirmation in China from Xi Jinping, who went to the Duisburg terminal on a visit to Germany in 2014, and to the Chongqing end in January 2016. As we will see in Chapter 5, the route sparked other cities in inland China to develop similar services, and prefigured the promotion of the silk road economic belt concept from late 2013. So far, the train from Chongqing has dominated total China–Europe freight train trade volumes, accounting for some 80 percent of goods by value,[31] though this proportion will inevitably decline as other cities develop their China–Europe train services.

The second set of geographical linkages developed and promoted by the Chongqing government from around 2010 were towards the east. These plans aimed to improve connectivity between Chongqing and traditional ports and coastal markets in the Yangzi River Delta, and further south to the Pearl River Delta. The aim here was less to encourage exports, though this is of course possible, but to integrate Chongqing more effectively with markets and production bases in the more developed regions of China. The route along the Yangzi was to be developed not simply through shipping as planned from the late 1990s, but also by the construction of road and rail linkages. This involved some challenging and expensive engineering. The link between Chongqing's main city and Wanzhou (the second city within the Chongqing municipality) – planned for completion by the end of 2016 – should reduce the 247 kilometre train journey to one hour. The section from Wanzhou to Yichang in Hubei opened in July 2014 and was one of the most challenging sections to construct, including 159 tunnels and 253 bridges.[32] These developments are complementary to the building of the Yangzi river economic belt, one of the three main regional strategies pushed by Li Keqiang's government since 2013. Of particular interest in Chongqing was the potential for this policy framework to promote further infrastructure connectivity to its east, and open up new economic and commercial opportunities.

The third geographical strand to Chongqing's global positioning is towards the southwest through Yunnan to connect to the economic and logistics corridors into southeast and south Asia. Three routes were planned to be operational by the end of 2016, from Chongqing through Guangxi to Hanoi, from Chongqing through Yunnan to Laos, and a route to Yangon in Myanmar.[33] The potential for trade creation or at least trade diversion here is possibly greater than that along the central Asian routes: China–ASEAN trade reached USD444 billion in 2013, with a target set of USD1 trillion by 2020. Most of this will still be carried out from the major trading provinces along China's coast, but greater connectivity through southwest China and to Chongqing increases the potential for Chongqing businesses to transport heavier industrial products to growing economies such as that in Vietnam. The Chongqing government has argued that all of these routes offer export potential, with the route through Kunming and Myanmar to connect to ocean routes to the Netherlands at 30 days, the China–Europe train taking 12 days, via Shenzhen 27 days and the longer-standing route through Shanghai over 40 days.

Setting out these routes demonstrated a striking global vision on the part of the Chongqing leadership, led at the time by Bo Xilai as municipal party secretary and Huang Qifan as mayor. The vision was reflected in Chongqing's twelfth five-year programme (2011–2015), with accompanying maps that place Chongqing at the centre of logistics connectivity across these three routes.[34] Delivering on the vision has required no less ambition, and here the main achievement has been to make the 'Yuxinou' Chongqing-Europe trains a regular feature of the logistics landscape for trade in goods to and from China. The maps produced by the Chongqing government are not otherwise dissimilar from those that have subsequently been produced to represent the belt and road initiative.

These developments, particularly those along the Yangzi river economic belt and to the southwest, also enhance connectivity with neighbouring parts of China, and therefore relate to two other contemporary regional policy constructs relevant to Chongqing's development in this era of globalization, the Chengdu–Chongqing economic belt and urban cluster and Yangzi river economic belt.

The Chengdu–Chongqing urban cluster and Yangzi River economic belt

Chongqing was of course not the only city in the central-western part of China looking to develop its economy and radically to upgrade its global economic and commercial ties. The most significant other city is Chengdu, some three hundred kilometres to Chongqing's west, whose policy development and status has reflected many similar themes to those in Chongqing. Indeed, Chengdu has been developing its own freight rail routes to Europe, following Chongqing's lead. The relationship between the two cities has been complex at least since the establishment of the PRC, and influenced the development paths of both. The primary relationship has been one of competition, but from the mid-2000s policy makers began to explore ideas of developing a Chengdu–Chongqing (or Chongqing–Chengdu) economic belt of up to 200,000 square kilometres and encompassing a population approaching 100 million. This would build on the way that the two cities were growing closer together as transport links developed following the upgrading of the road and rail connections and the launch of high-speed train services between the two. From a policy perspective when seen at the national level, the aim was further integration of the two city economies, building on their areas of complementarity, and driving economic growth in that part of western China. Both cities, however, continued to treat the relationship primarily as a competitive one and seek their own policy advantages. Nonetheless in April 2007 – when the open-minded Wang Yang was party secretary in Chongqing – agreements at the provincial level between Chongqing and Sichuan made progress towards establishing the Chengdu–Chongqing region as a concept that would drive economic policy.[35]

In 2011 an official plan for a Chengdu–Chongqing economic zone/region was published, followed in April 2016 by a development programme for a Chengdu–Chongqing urban cluster.[36] This plan was not only prompted by debates at the provincial level in Chongqing and Sichuan/Chengdu, but built on national spatial policy programmes developed over the preceding years, in particular those on functional zones (2010), new-type urbanization (2014) and the Yangzi River economic belt (2014), all of which were reflected in the 13th five-year programme covering 2016–2020. The April 2016 programme defines the extent of the urban cluster, to cover 27 of Chongqing's 40 districts (or counties) and 15 cities in Sichuan, with total area of 185,000 square kilometres, or 1.9 percent of China's land mass. In 2014, the resident population was 90.94 million and GDP of the cluster was RMB 3.76 trillion, accounting for 6.6 percent and 5.5 percent of the national totals respectively. This creates substantial market potential, as there are populations of 200 million within a 500 kilometre radius and 500 million within a 1,000 kilometre radius of Chongqing.

The overall goal of these plans is to encourage more coordination and less duplication between the areas around Chengdu and those in Chongqing, pushing against the tendencies of administrative region economies and the bureaucratic competition between the two cities. To deliver this, the programme sets out key projects from transport to industry and environmental protection, the industrial and economic focus for various cities and districts within the cluster, and how the spatial structure and logistics and infrastructure connectivity should operate. There is a significant emphasis on the cluster developing external openness, both within China to other provinces and regions, and internationally. A map indicating the global connections shows six major arrows, respectively along the Yangzi to the ocean, southeast to Guangdong and beyond, northeast to the Beijing–Tianjin–Hebei cluster, northwest through Xinjiang, west through Tibet, and southwest through Yunnan, all connecting on to 'international trade corridors'.

How this cluster and the relationship between Chongqing and Chengdu will develop remains to be seen. Policy goals aside, the growth of infrastructure connectivity between the two cities, and the appeal to businesses of the market scale of an economic region encompassing both will drive greater integration. On the other hand, the sense of competition between the two cities is deep seated, and duplication of projects and competition to attract outside investment can be expected to continue. Indeed, there has been more enthusiasm in Chongqing over the last couple of years in engaging with the Yangzi river economic belt (YREB) concept than the Chengdu–Chongqing economic cluster (though the two are linked). This is partly the result of historical orientations, the competitive relationship between Chongqing and Chengdu, and Chongqing's long-standing desire to develop and economic and industrial corridor along the Yangzi, dating back to the 1990s and cooperation agreements with Wuhan, Nanjing, and Shanghai.[37] Given that the YREB and the belt and road initiative are two of the priority regional policy initiatives of the central government after 2013, the Chongqing government has seen particular benefit to be gained from highlighting its position at the intersection of the YREB and the silk road economic belt, as well as the routes from Chongqing to the southwest and into southeast Asia. These developments demonstrate not just a domestic focus, but the way that regional actors in China perceive themselves in global context.

Bo Xilai in Chongqing: challenging globalization?

However, the idea that Chongqing's approach to development from 2007 to 2012 reflected a desire to engage with globalization might be questioned by some who view the policy approaches of Bo Xilai, Chongqing's controversial party secretary for over four years, in terms of a revival of a socialist orientation. It was Bo's tenure in Chongqing, and especially his demise, which brought the city to the forefront of international attention for much of 2012. The demise began in February 2012 with news that Chongqing's former chief of police, Wang Lijun, had fled to the US consulate-general in Chengdu. Wang had been the right-hand man of Bo Xilai, the charismatic princeling and member of the 25-strong Politburo at the top of the Party's national hierarchy, who had been sent somewhat reluctantly to Chongqing as municipal party secretary in November 2007. Wang was quickly sent from the consulate to Beijing, and after an awkward hiatus, Bo was removed from his post in Chongqing in the middle of March, immediately after the conclusion of the annual meetings of the people's congresses (legislative and advisory bodies) in Beijing. A month later he was suspended from the Party's Politburo and Central Committee and investigated for serious violations of Party discipline. He was expelled from the Party at the end of September 2012 (just before the national leadership transition), and handed to the judicial apparatus for a high-profile trial that took place in August 2013. In an enigmatic part of the story, Bo's wife was found guilty of the murder in November 2011 of a British business man in Chongqing. In March 2012 Bo was replaced as municipal party secretary by Zhang Dejiang, an experienced Politburo member and vice premier. When Zhang joined the seven-strong Politburo Standing Committee in November 2012, Sun Zhengcai was sent to Chongqing as party secretary, though Sun himself was removed from his post in July 2017 and detained for alleged corruption.

Beyond these basic facts, little about the end of Bo's time in Chongqing is certain, and contrary to the impression given in some accounts,[38] most of us will probably never know precisely what happened. Because it erupted in advance of a major leadership transition at the top of the Communist Party (which took place on schedule in November 2012), one interpretation has been that Bo's demise was the result of infighting at the top of the Party, or had perhaps even been contrived by Xi Jinping, soon to take over from Hu Jintao as Party general secretary, to remove Bo. For this writer, however, hubris and human failings on the part of Bo and his wife have always seemed more plausible explanations.

The case sparked debate for another reason, which is more relevant to the themes of this book, Bo's policy record while serving in Chongqing. Bo had first made his mark in the city through public negotiations with striking taxi drivers and a high-profile and broadly-popular 'strike black' (*dahei*) crackdown on corruption and cartels in the city, as well as the 'umbrella' of officials who had sheltered them. Some saw this as an effort to target the legacy of his predecessor, Wang Yang, who had moved to run Guangdong province just before Bo arrived in Chonging.

In addition to Bo's authoritarian-tinged law and order campaign, there were suggestions that he might have been experimenting with or promoting a more 'socialist' alternative to the Party's mainstream policy approach. Some of Bo's comments in Chongqing encouraged this way of thinking. He talked, for example, of 'common prosperity', a phrase used by Deng Xiaoping in December 1990 to describe the greatest advantage of socialism, a somewhat different emphasis from Deng's earlier injunction that some should 'get rich first'.[39] As well as saying that the city's living environment should be improved, Bo set a policy goal of reducing the Gini coefficient measuring income disparities in Chongqing to 0.35 by 2015, at a time when the central government was reluctant to publish Gini coefficients for the country as a whole. He enthusiastically promoted the policy of building 'social housing', and encouraged Chongqing cadres to spend time in rural areas and 'get close to the masses'. This sounded as if it echoed Party policies during the Mao era, and – in particular in the lead up to the 90th anniversary of the Communist Party's establishment on 1 July 2011[40] – Bo encouraged nostalgia through the 'sing red' (唱红 changhong) campaign to recover songs from the Party's earlier years (Bo's contrasting 'black' and 'red' campaigns, reminiscent of cultural revolution divisions into black and red elements, was probably no accident). Major 'red song' gatherings were organized in Chongqing, one of which was even attended by Henry Kissinger. Chongqing's bookshops featured titles such as *Learn strategy from Mao Zedong* and *Learn management from Mao Zedong*.[41] Bo also engaged in some local identity politics, such as holding a high-profile celebration in November 2009 of the 60th anniversary of Chongqing's 'liberation', which had not taken place until after the formal establishment of the PRC on 1st October 1949.

Bo's policy stances attracted interest from a number of China's 'new left' academics, who had since the 1990s been challenging the Party's approach for being too neoliberal or capitalist. In 2011, a special issue of the academic journal *Modern China* was devoted to Chongqing, with its editor Philip Huang arguing that in Chongqing 'equitable development' was driven by a 'third hand' between the market and state; an article in that journal by one of China's new left intellectuals, Cui Zhiyuan, talked about an 'effort [in Chongqing] to revitalize the party's relationship with the people';[42] and Chongqing's approach under Bo was characterised by influential Chinese academic Wang Shaoguang as 'socialism 3.0'.[43] As early as 2009, there was some talk of the development of a 'Chongqing model'. As mayor, Huang Qifan talked instead of 'new economic policies' in an extensive interview in Caixin's *China Reform* magazine of November 2010, and the Chongqing leadership consistently rejected the idea of a 'Chongqing model', with Huang doing so prominently again after Bo's removal from Chongqing.[44]

Whether to characterise these as experiments, innovations, or challenges to national policy, and how much socialism to infer from the policy approach, are two of the analytical questions to be addressed in evaluating Bo's time in Chongqing. They have generally been seen as challenges, part of the wider political narrative in which Bo's 'rise and fall' was driven by a challenge for national leadership. Here Bo's approach of 'sharing the economic cake more fairly' is contrasted with that of 'growing the cake', and his policies in Chongqing compared to the more liberal economic approach of Wang Yang in Guangdong.[45] But although Bo did seem at the time to be rather less collegiate than might have been expected of Politburo members under Hu Jintao's collective leadership, the contrast between approaches in Guangdong and Chongqing can also be seen as part of the process of sub-national experimentation and testing of policy approaches, which has been a mainstream feature of China's policy making processes throughout the PRC's history. As Politburo members running provinces/municipalities away from Beijing and Shanghai, both Bo and Wang had the space to test ideas and approaches without this necessarily being politically subversive. Their approaches responded to the different political economies of the two municipalities, and the need for 'complimentary accumulation strategies' that reflect 'the different positionalities of each province in the uneven, regionally variegated pattern by which China has become progressively integrated in global capitalist production networks'.[46]

As for the apparent 'socialist' tilt, further examination of Bo's policies, and the policies discussed above developed under him by Huang, reveals some underlying contradictions in this analysis – though the *perception* that Bo's demise closed off an alternative model has its own implications for official ideology, and happened to coincide with some debate over China's future direction sparked by the publication in February 2012 of the World Bank's *China 2030* report, which advocated further market-oriented reforms. But national-level policies were actually reflected in many of the measures promoted by Bo. For example, social housing was fast becoming a national-level policy priority, with a new target promulgated in March 2011 and included in the twelfth five-year programme. Even though the national-level Gini coefficient remained sensitive, the goals of reducing income inequality and engaging in more proactive income redistribution had been moving up the national policy agenda through the 2000s. And for all Bo's talk about sharing the economic cake more fairly rather than 'growing' the cake, Chongqing's GDP growth placed it at the forefront of provincial economies during his tenure, up 14.9 percent in 2009, 17.0 percent in 2010, and 16.4 percent in 2011, even after the global financial crisis had hit national GDP growth levels hard. It can further be debated whether the emergence under government guidance of major state-owned conglomerates – the 'eight big groups' as they were known in Chongqing – was a manifestation of a socialist tilt, or another step in the development of something that is better described as 'state capitalism'. Along with Shanghai, the municipality was chosen to trial a property tax, and pilot measures to promote coordinated development of rural and urban areas, initiated under Bo's predecessor, but carried through during his tenure. Neither of these key elements, or the inauguration of the Liangjiang New Area in 2010, or tax-efficient bonded customs zones, hint at socialism.

In fact, Bo's time in Chongqing saw the government placing great emphasis on the development of foreign trade and investment, the area Bo had been responsible for in his previous job as minister of commerce from 2004 to 2007. Chongqing's trade and investment volumes grew rapidly. Bo focused particularly on outward investment by Chongqing companies, at one point setting a hugely ambitious target of USD 30 billion over five years. One of the signature slogans for Chongqing from Bo's time was to make it a 'high point for openness in inland China' (language that has since been echoed by other provinces and in the belt and road initiative). All this smacks of engagement with globalization much more than a red-tinged socialist turn. In fact, rather than challenge national strategies, if anything Chongqing's experience has been basis for subsequent policy development, as the China–Europe trains and the development of the silk road concepts demonstrate.[47]

Conclusion

A period of some political uncertainty following the demise of Bo Xilai was at least temporarily put to an end when Xi Jinping visited the city at the beginning of January 2016 and affirmed Chongqing's approach to economic development. He visited the Chongqing end of the 'Yuxinou' China–Europe train route, a symbol of the belt and road initiative. He also went to Guoyuan port, again symbolic as a link between the silk road economic belt and connectivity towards Chongqing's east through the Yangzi river economic belt. These elements came together in Xi's injunction that the municipality should develop into an international logistics hub.[48] This is echoed in the Chengdu–Chongqing urban cluster programme, and the positioning of Chongqing and Chengdu in the Yangzi river economic belt initiative.

In other words, the message is that Chongqing should continue to cement its position as a key node in networks across the domestic and global economy, through ongoing integration into global production networks in particular sectors, and by further developing domestic and global transport projects to overcome the 'tyranny of distance' it faces as an inland city far from external and many domestic markets. This distance can be overcome by the application of technology and the ways in which the global economy has developed, including through informationalization of the knowledge economy. Since the 2000s, Chongqing – led by government at the municipal level – has been a particularly interesting, confident and innovative example of a region of China trying to fit into and respond to capitalist globalization and China's integration into it. In doing so, as can be seen by the government's engagement with global businesses, the practices this has led to reflect an integration into and acceptance of the structures of the global capitalist economy as they are, albeit with the goal of enhancing Chongqing's – and China's – status within it, and therefore ability to benefit from the further development of China's relationships to the global economy. In sum, Chongqing's development shows that globalization, and China's engagement with it, has been moving inland.

Notes

1 As well as the primary and secondary sources cited in the text, this chapter draws on the author's posting from 2004 to 2007 as British consul general in Chongqing, and on regular visits back to the city since 2007.
2 Coates, P. D. *China consuls: British consular officers, 1843–1943*. Oxford, UK: Oxford University Press, 1988.
3 Graham Peck. *Two kinds of time*. Cambridge, MA: Houghton Mifflin, 1950.
4 Hong Lijian. 'A tale of two cities: a comparative study of the political and economic development in Chengdu and Chongqing'. In Jae Ho Chung, ed. *Cities in China: recipes for economic development in the reform era*. London and New York: Routledge, 1999, pp. 189–191.
5 Hong, 'A tale of two cities' (see note 4), pp. 193–195.
6 Kjeld Allan Larsen. *Regional policy of China, 1949–85*. Manila: Journal of Contemporary Asia Publishers, 1992, p. 94.
7 Caixin. 'Chongqing mayor says rural land reform pilot has been just the ticket', 17 September 2015, www.caixinglobal.com/2015-09-17/101012190.html.
8 Fung Business Intelligence. 'Chongqing: Changjiang shangyou de zhongxin chengshi' ('Chongqing: a central city on the upper reaches of the Yangzi'), 2016, www.fbicgroup.com/sites/default/files/Clusters_Series_special_Chongqing.pdf, slide 24.
9 In July 2013, Mayor Huang Qifan declared Chongqing's ambition to become the Detroit of China, though Detroit itself was not faring well at the time. See: Sohu.com. 'Huang Qifan: Chongqing jiang chengwei Zhongguo de Ditelv' ('Huang Qifan: Chongqing will become China's Detroit'), 19 July 2013, http://business.sohu.com/20130719/n382067427.shtml.
10 Edward Steinfeld. *Playing our game: why China's rise doesn't threaten the West*. Oxford, UK: Oxford University Press, 2010, p. 107.
11 Economist Intelligence Unit. *China hand: the complete guide to doing business in China 2014*, www.eiu.com, 2014, p. 100.
12 Dali Yang. 'China adjusts to the world economy: the political economy of China's coastal development strategy'. *Pacific Affairs*, 64: 1, 1991, p. 51. See Chapter 2 of the present book for further discussion.
13 Caixin. 'Chongqing "xin jingji zhengce" ' ('Chongqing's "new economic policies" '). *Zhongguo gaige* [*China Reform*], 325, November 2010, pp. 14–16.
14 Steinfeld, *Playing our game* (see note 10), pp. 97–99.
15 Chongqing News. 'Huang Qifan: "shanzhai qiye" neixuxing jiagong maoyi moshi zhide jiejian' ('Huang Qifan: the domestic demand trade processing model of "copy-cat enterprises" is worth learning from'), 3 July 2009, http://big5.news.cn/gate/big5/cq.news.cn/2009-07/03/content_16986860.htm.
16 Wen Wei Pao. 'Huang Qifan zhili Chongqing wu da "qi" zhao' ('Five "rare" moves in Huang Qifan's governance of Chongqing'), 4 August 2015, http://news.wenweipo.com/2015/08/04/IN1508040026.htm.
17 Caixin, 'Chongqing mayor' (see note 7).
18 Chongqing Customs. 'Huang Qifan tuijie Chongqing shangji: shenchu neilu, que yidian bu bi yanhai cha' ('Huang Qifan promotes Chongqing business opportunities: location in inland China but no worse than the coast'), 25 February 2011, http://chongqing.customs.gov.cn/publish/portal153/tab34627/module76185/info324619.htm.
19 This will be discussed further in the conclusion to this book (Chapter 6).
20 Fung Business Intelligence, 'Chongqing' (see note 8), slides 17–23.
21 STCN, 'Chongqing 2017 nian fuwu maoyi jinchukou zong'e jiang da 300 yi meiyuan' ('In 2017 Chongqing's total trade in services will reach USD 30 billion'), 12 August 2015, www.stcn.com/2015/0812/12406602.shtml.
22 Chongqing Foreign Trade and Economic Relations Commission. '2014 nian Chongqing shi fuwu maoyi yunxing jiankuang' ('Basic situation of Chongqing's trade in services in 2014'), 2014, www.ft.cq.cn/zxfw/ljcqfwmy/.

23 Liangjiang means two rivers, a reference to the Yangzi and Jialing, at whose confluence the city of Chongqing sits. Since 2010, new areas have been approved in other western provinces, including Guizhou, Shaanxi, Gansu and Sichuan (the Tianfu New Zone centred on Chengdu, established in 2014). The most recent new area, in Xiong'an, Hebei province, has attracted most attention (see Chapter 2).
24 Chongqing Liangjiang New Area Management Committee, 'Chongqing Liangjiang new area'. No date. Author's personal archive.
25 Wen Wei Pao, 'Huang Qifan' (see note 16); Dinny McMahon. 'The terrible amusement park that explains Chongqing's economic miracle.' 29 August 2016, *Foreign Policy*. http://foreignpolicy.com/2016/08/29/chongqing-economic-miracle-locajoy-debt-sales-state-owned-enterprises/.
26 Barry Naughton. 'Supply-side structural reform at mid-year: compliance, initiative, and unintended consequence'. *China Leadership Monitor*, 51, 2016, www.hoover.org/research/supply-side-structural-reform-mid-year-compliance-initiative-and-unintended-consequences.
27 *Xinhua*. 'Zhonghuo renmin gongheguo guomin jingji he shehui fazhan di shisan ge wunian guihua gangyao' (Outline of the 13th five-year programme for national economic and social development of the People's Republic of China), 17 March 2016, http://news.xinhuanet.com/politics/2016lh/2016-03/17/c_1118366322.htm.
28 There are various figures cited for the duration of train journey to Europe, from 12 days up to 17 days.
29 Data collated from a range of Chinese media sources and the website of Yuxinou (Chongqing) Logistics Co. Ltd. (http://yuxinoulogistics.com), as well as Fung Business Intelligence, 'Chongqing' (see note 8), slides 28–29.
30 DHL, 'Enhanced China-Kazakhstan-CIS-Europe connectivity: DHL and KTZ Express sign MOU to enhance China-Kazakhstan rail freight link', 30 June 2015, www.cn.dhl.com/en/press/releases/releases_2015/local/063015.html.
31 Fung Business Intelligence, 'Chongqing' (see note 8), slide 29.
32 *Xinhua*. 'Bullet train service starts on China's most challenging railway', 1 July 2014, http://news.xinhuanet.com/english/china/2014-07/01/c_133453059.htm.
33 *Xinhua*. 'China exclusive: Highway corridor links Chongqing to SE Asia', 4 August 2016, http://news.xinhuanet.com/english/2016-08/04/c_135564313.htm.
34 A similar map was available on the website of the Invest in Chongqing agency, at www.investincq.com/public/news/detail/2082.html (accessed 6 October 2016).
35 Tim Summers. 'China's western regions 2020: their national and global implications'. In Kerry Brown, ed. *China 2020: what policy makers need to know about the new rising power in the coming decade*. Oxford, UK: Chandos, 2011, pp. 137–171. See Hong, 'A tale of two cities' (see note 4) for historical background.
36 National Development and Reform Commission. 'Cheng-Yu chengshiqun fazhan guihua' ('Development programme for the Chengdu–Chongqing urban cluster'), April 2016, www.ndrc.gov.cn/zcfb/zcfbghwb/201605/W020160504587323437573.pdf. Three other city cluster programmes were promulgated at around the same time, covering cities in northeast China (Harbin-Changchun), the middle reaches of the Yangzi, and the Yangzi River Delta.
37 Hong, 'A tale of two cities' (see note 4), p. 203.
38 John Garnaud. *The rise and fall of the House of Bo*. Melbourne: Penguin, 2012; Jamil Anderlini. *The Bo Xilai scandal: power, death, and politics in China*. London: The Financial Times, 2012.
39 'Socialism is not a small number of people getting rich and the vast majority remaining poor, it is not like that. The greatest advantage of socialism is common prosperity, that is the thing which embodies the nature of socialism'; cited in: Hu Angang, Yan Yilong, and Wei Xing. *2030 Zhongguo: maixiang gongtong fuyu [2030 China: towards common prosperity]* Beijing: People's University Press, 2011, p. 129 (author's translation). Deng also said in 1992 that 'Socialism's real nature is to liberate the productive forces, and the ultimate goal of socialism is to achieve common prosperity', though as Maurice Meisner

notes, this could said about the nature of capitalism as much as of socialism; Maurice Meisner, *Mao's China and after*. New York: Free Press, 1999, pp. 489–490.
40 The lack of top leadership enthusiasm for a Maoist turn was one of the messages delivered in July 2011 around the 90th anniversary of the Party's founding.
41 This author picked up both volumes on a trip to Chongqing during Bo's time in charge. As well as Mao, their authors cite Western 'gurus' such as Jack Welch.
42 Cui Zhiyuan. 'Partial intimations of the coming whole: the Chongqing experiment in light of the theories of Henry George, James Meade, and Antonio Gramsci'. *Modern China*, 37: 6, 2011, p. 658.
43 Wang Shaoguang. 'Chinese socialism 3.0', translated in Mark Leonard, ed. *China 3.0*. European Council on Foreign Relations, 2012, pp. 60–67, www.ecfr.eu/page/-/ECFR66_CHINA_30_final.pdf; Global Times. 'Chongqing pioneers China's "socialism 3.0"', 24 September 2010, www.globaltimes.cn/content/576673.shtml.
44 For the 'Chongqing model' idea see: Chinanews.com. 'Yazhou Zhoukan: Chongqing moshi chuang Zhongguo jingji fangong xin lujing' ('Asiaweek: Chongqing model provides a new path for the Chinese economy'), 6 February 2009, www.chinanews.com/hb/news/2009/02-06/1552353.shtml; for Huang's 2010 interview see: Caixin, 'Chongqing's "new economic policies"' (see note 13). For subsequent repudiation of the model, see: China.com.cn. 'Chongqing daibiaotuan kaifangri: Bo Xilai Huang Qifan da Zhongwai jizhe tiwen' ('Chongqing delegation's open day: Bo Xilai and Huang Qifan answer questions from Chinese and foreign journalists'), 8 March 2011, www.china.com.cn/2011/2011-03/08/content_22085020.htm.
45 See the discussion of Wang's time in Guangdong in Chapter 3 of the present book. In a May 2014 article in the Party's theoretical journal *Qiushi* (*Seeking truth*), premier Li Keqiang identified a need to grow the cake and to cut it better.
46 Andreas Mulvad. 'Competing hegemonic projects within China's variegated capitalism: "liberal" Guangdong vs. "statist" Chongqing'. *New Political Economy*, 20, 2015, p. 208.
47 There has since been some debate around the idea that Xi Jinping has indeed delivered a Maoist turn. I remain highly skeptical: as historian of China under Mao Andrew Walder said in an interview with the Asia Society in May 2015, 'Xi is a conservative nationalist. Mao was a radical and a revolutionary. Xi wants stability above all; Mao saw disorder and conflict as the only way to move forward. So the parallels are entirely superficial.' (see http://asiasociety.org/blog/asia/interview-how-frustrated-ideologue-mao-zedong's-failing-bureaucracy-derailed-china).
48 *Xinhua*. 'Xi envisions Chongqing as int'l logistics hub, stressed development', 7 January 2016, http://news.xinhuanet.com/english/2016-01/07/c_134984365.htm.

5
THE BELT AND ROAD INITIATIVE AND CHINA'S REGIONS

The growing global engagement and ambitions of Chongqing are not relevant just for Chongqing, but are a precursor to one of the main innovations in Chinese policy under Xi Jinping, the 'belt and road initiative', often referred to as 'one belt, one road'.[1] This initiative has generated plenty of debate already,[2] but most of this has analysed the initiative for what it tells us about China's global diplomatic or economic strategy. However, it is important also to examine it from the perspective of China's regions, not least given the regional emphasis in the central government's own policy statements on the belt and road initiative. Doing so shows that there are important regional origins to the belt and road initiative, building on earlier evolutions in China's regional policy. Further, the implementation of the initiative will be in a large part dependent on how various provinces respond to the belt and road initiative, which in turn has implications for regional policy in China and the development of China's regions in an era of globalization.

The belt and road initiative and China's regions

China's belt and road initiative is an omnibus policy framework, the stated aims of which are to enhance connectivity from Asia through Africa and the Middle East to Europe. It consists of two complementary concepts, which were set out in autumn 2013 during overseas trips by China's president Xi Jinping and premier Li Keqiang. The 'belt' refers to the proposal to build a 'silk road economic belt' across the Eurasian land mass, while the 'road' is a reference to a '21st-century maritime silk road', primarily passing through the South China Sea and Indian Ocean.[3] Neither concept should be understood as being limited to a singular route or to locations that can be clearly marked on a map, but rather as complex and dynamic networks of connectivity (see Map at the front of this book). Although earlier assessments by Chinese sources suggested that the initiative covered

65 countries,[4] the Chinese government has stated that it is open to the 'active participation of all countries', and there is no official list of countries covered by the initiative. In a high-level meeting in Beijing in August 2016, Xi Jinping said that 'more than 100 countries and international organizations have participated in the Belt and Road Initiative',[5] a message reiterated in the high-profile Belt and Road Forum hosted in Beijing in May 2017.

The key reference for Chinese policy on the belt and road initiative is a 'vision and actions document' issued by the State Council at the end of March 2015.[6] As well as the material on China's regions discussed in detail below, this document sets out the principles and approach proposed by the Chinese government, the institutional mechanisms for the initiative's implementation (including numerous multilateral institutions), and details of the sort of connectivity envisaged as part of the belt and road. The initiative reflects numerous themes in Chinese foreign policy,[7] and the vision and action document was promulgated shortly after Xi Jinping gave a speech setting out a vision for Asia at an annual international forum in Boao, Hainan island. As well as reflecting foreign policy goals, the initiative should be understood as presaging the further incorporation of China into a global capitalist economy, on terms set more by China than in the past.[8] The initiative has already been widely debated, and the broad omnibus nature of the vision and actions document almost inevitably allows for multiple interpretations both of Chinese intentions and of the implications of the initiative. One motivation for the belt (though not so much the maritime 'road') appears to be a way of stimulating further economic development in some of the less developed and less open inland provinces, building on this aspect of regional policy developed in the first part of the 21st century.

Further messages about the objectives of the belt and road initiative can be deduced from the two places it is covered in the thirteenth five-year programme, the document that sets out the government's social and economic development strategy for 2016–2020.[9] The first reference is in Chapter 9 of the programme, on 'coordinated regional development', where 'one belt, one road' is mentioned as the first of the government's three regional policy initiatives and as a useful driver for the development of western China. The second reference is in Chapter 11 of the 13th five-year programme covering 'comprehensive openness', where the belt and road is described as 'overseeing' (*tongling*) the development of a new level of China's openness. A section in the chapter is devoted to the belt and road, and identifies three key priorities: the building of belt and road cooperation mechanisms, promoting open economic corridors,[10] and enhancing open and inclusive people-to-people cultural exchanges.

The relevance of regional perspectives to the belt and road initiative can also be seen directly from the 2015 vision and actions document itself. Section VI of the document is entitled 'China's regions in pursuing opening-up'. The section is light on explanation and instead sets out a list of the roles that particular provinces and cities in China are expected to play in the initiative. The principles on which this is based are stated as being to 'leverage the comparative advantage of [China's] various regions, adopt a proactive strategy of further opening-up, strengthen interaction and cooperation among the eastern, western and central regions, and comprehensively improve the openness of the Chinese economy'. These principles are familiar elements of contemporary regional policy in China, including the point about comparative advantage, which has been a guiding principle since the 1980s.

However, rather than follow the demarcation of China into eastern, western and central regions hinted at by these chapeau principles, Section VI is structured around four broad macro-regions along lines not seen in earlier statements of regional policy. These are identified as the northwestern and northeastern regions; the southwestern region; coastal regions and Hong Kong, Macao and Taiwan; and inland regions. The document does not give explicit definitions of these regions but instead identifies relevant policy goals for various provinces, major cities and economic zones that fall within each mega-region.

The document demarcates the northwest as Xinjiang, Shaanxi, Gansu, Ningxia, and Qinghai (in that order). A number of points are made about the way in which Xinjiang could become a 'core area on the Silk Road Economic Belt' (see below), whereas the other provinces are only noted briefly for their 'economic and cultural strengths' (Shaanxi and Gansu) and 'ethnic and cultural advantages' (Ningxia and Qinghai) – for example, Ningxia has been positioning itself as a link to the Muslim world given its own Hui (Muslim) minority status, including by hosting a China–Arab States Expo, trying to attract investment from Arab countries, and developing its local halal food industry.[11] Three cities are mentioned by name, Xi'an, Lanzhou, and Xining, along with the Ningxia Inland Opening-up Pilot Economic Zone. For the northeast, the scope of the policies is familiar from the evolution of the post-2003 'revive the northeast' policy framework, based on making use of Inner Mongolia's proximity to Russia and Mongolia and Heilongjiang's to Russia, while the traditional three northeast provinces of Heilongjiang, Jilin, and Liaoning are encouraged to strengthen cooperation and linkages with Russia's far east.

The second macro-region covered is the southwest. This contains material on Guangxi and Yunnan (see below), as well as the idea of promoting 'border trade and tourism and culture cooperation between Tibet Autonomous Region and neighbouring countries such as Nepal'. This marks something of a policy departure: in the 12th five-year programme, Tibet was not identified as a region that should engage in transborder interactions and opening up to its land neighbours. This seems to have changed under the belt and road initiative, and there has since been active discussion within China about the sort of role that Tibet might play, including through the development of border trade, international tourism, and in supporting industries in Tibet such as Tibetan medicine and animal husbandry. There have also been discussions about new infrastructure connections between Tibet and Nepal, including rail linkages. Potential sensitivities in India may have led a role for Tibet, particularly in linking to South Asia beyond Nepal, to have been played down, but it does seem that the belt and road has sparked a new phase in

Chinese thinking about Tibet's external linkages. At this stage, however, it is not clear how a role for Tibet might further develop.[12]

The third macro-region – coastal regions – is dealt with in detail. Here the spatial structures identified as playing a role in the belt and road are not provinces, but a series of other regional constructs, specific zones and cities, reflecting nodes in the spatial networks that have become a feature of this era of globalization.[13] The ideas can be grouped under a number of themes:

- An aim to 'leverage the strengths' of strong and open economic regions along the coast, in particular the Yangzi River Delta, the Pearl River Delta, the west coast of the Taiwan straits and the Bohai Rim, as well as speeding up development of the free trade zone (FTZ) in Shanghai,[14] and supporting Fujian as a 'core area' of the maritime silk road.
- Further focus along the southeast coastal area on the three experimental zones of Qianhai, Hengqin and Nansha in Guangdong, and Pingtan in Fujian (since upgraded to FTZs), and building the 'Guangdong–Hong Kong–Macao Big [Greater] Bay Area'.
- Developing maritime-focused zones in Zhejiang and Fujian, and Hainan island's international tourism.
- Strengthening port construction in a series of cities: Shanghai, Tianjin, Ningbo-Zhoushan, Guangzhou, Shenzhen, Zhanjiang, Shantou, Qingdao, Yantai, Dalian, Fuzhou, Xiamen, Quanzhou, Haikou, and Sanya. Strengthening international hub airports in Shanghai and Guangzhou.
- Encouraging overseas Chinese, Hong Kong, and Macao to participate and contribute to the initiative, and making 'proper arrangements for the Taiwan region to be part of this effort' (see the section on Fujian below for some further discussion of Taiwan).

As with the coastal section, the material on the final macro-region – inland regions – does not focus on provinces but on city clusters: the Chengdu–Chongqing area, central Henan province, Inner Mongolia around Hohhot, Baotou, Erdos, and Yulin, and around Harbin and Changchun (these cities are in northeast China, which already featured as part of the first macro-region). There are some further city goals: Chongqing should be 'an important pivot for developing and opening up the western region', and Chengdu, Zhengzhou, Wuhan, Changsha, Nanchang, and Hefei 'leading areas of opening-up in the inland regions'. There follow references to cooperation between regions on the upper and middle Yangzi and 'counterparts along Russia's Volga River', coordination along the China–Europe corridor and 'cultivat[ing] the brand' of 'China–Europe freight trains' (see below), constructing a cross-border transport corridor that would connect the eastern, central and western regions, and supporting airport and international land port development in cities such as Zhengzhou and Xi'an.

This is a broad and high-level series of proposals, each of which have their own historical trajectory. Other than in western China, the emphasis on trans-jurisdictional regions and municipalities raises questions about implementation, given the strength of administrative units in economic development and the challenges experienced in the past in delivering policy goals that cut across provincial boundaries.[15] It may be that much of the implementation falls back on provincial units as well as cities, as this is where the resources and bureaucracies lie, though at this stage it seems misleading to identify the regional focus solely in terms of provinces, as has been done in some reports.[16]

Finally in this section, it is worth noting that there appear to be several priority provinces under the initiative. The vision and actions document identifies two provincial units, Xinjiang (an autonomous region) and Fujian (province), with the status of 'core areas' of the belt and road respectively, and the description of Xinjiang's role is particularly detailed. At a major Party-state work forum to pursue the construction of the belt and road chaired by Xi Jinping in Beijing on 17 August 2016, the Party secretaries from both Xinjiang and Fujian, as well as Guangdong and Shaanxi, were key speakers, giving their provinces greater domestic profile in the implementation of the initiative.[17] Shaanxi's capital, Xi'an was – with its historical name of Chang'an – the start of the historical silk road routes across Eurasia, and Shaanxi may also be a priority as Xi Jinping's 'home province'.[18] When it comes to Fujian (see below), there has been speculation that its prominence may be a product of the lengthy period time that Xi Jinping spent in the province earlier in his career (from 1985–2002), during which he was attracted by the ideas of developing a contemporary maritime silk road, and there has been much discussion over the years of the role in historical maritime trading links of ports such as the Fujian city of Quanzhou. To explore these ideas further, we turn to the question of regional origins of the belt and road initiative itself.

Regional origins of the belt and road initiative

As should be clear from this brief description, there are multiple motivations for the belt and road initiative, from considerations of foreign policy, to furthering global economic integration. In addition to these, looking at the initiative from the perspective of China's regions shows that the ideas set out by the Chinese leadership as the 'belt and road' do not represent fundamentally new policy content but the evolution of long-standing approaches to global interactions of at least some of China's regions.[19] This means that, even though the linkage is not explicitly acknowledged, it can be argued that provincial agency has been instrumental in creating the foundation on which the national-level silk road vision sits. This is particularly the case in two areas to be discussed in more detail in this section, the shift to open up China's land-locked western provinces to neighbouring economies to China's west, for which Xinjiang and Yunnan/Guangxi are good examples, and policies developed in Fujian and Guangdong that drew on historical maritime trading links for inspiration; all of these are cases in which provincial policy elites had previously utilized metaphors of the 'silk road' in developing policies to promote international interactions. Provincial agency also appears to be at work with respect to northeast China and the Russian Far East, where the silk road economic belt has effectively taken 'decades-old ideas and incorporated them into a new framework'.[20]

Xinjiang and central Asia

Existing research on Xinjiang has long identified strategies emerging from Xinjiang to build links with its Islamic neighbours and to become the 'nexus of a Silk Road economy in the Great Islamic Circle'.[21] Responding to the priority given to coastal regions in the early years of reform and opening up, policy makers in Xinjiang promoted a 'double opening' of Xinjiang from the 1980s, 'an attempt to turn Xinjiang's geopolitical position to China's advantage'.[22] This envisaged that opening to the region's east would lead to closer integration into the national economy, while to the west the region would develop economic ties with what was then still Soviet central Asia, following the reopening of border trade between Xinjiang and central Asia in 1983. The subsequent normalization of Sino-Soviet ties in the late 1980s made further opening possible and by 1990, a train line from the regional capital Urumqi to the Alatau Pass at the border with Kazakhstan had opened.[23] In 1986, the regional government set out a goal of 'comprehensive opening to the west', and in 1989 referred formally to 'double opening'. During this period, the silk road was 'one prominent metaphor used to depict Xinjiang's intermediate position between China, Central Asia and beyond', one that was particularly useful because it allowed discussion of cross-continental linkages while avoiding the sensitive question of whether Xinjiang could be considered as part of central Eurasia.[24]

However, Xinjiang's trajectory during the post-Cold War 1990s proved more complex. Soviet central Asia gave way to a series of new independent states, the 'Stans', three of which would become members of the Shanghai Five with China and Russia in 1996, and – joined by Uzbekistan – found the Shanghai Cooperation Organization (SCO) in 2001. In the context of a wave of post-Cold War self-determination, some in Xinjiang began to seek greater autonomy, religious and cultural freedom, or even independence. China's primary motivation for cooperation with central Asian states was dealing with transborder security concerns to meet what its government called the 'three evil forces' of separatism, extremism, and terrorism. In this context, the cross-border movement of people (essential for trade) had the potential to pose a security risk in Xinjiang.[25]

Nonetheless, Chinese policy makers at both the local and national level retained an interest in greater economic and commercial interactions between Xinjiang and central Asia. In 2003 premier Wen Jiabao proposed that the SCO might devise a free trade agreement,[26] though this idea was not taken up by the organization's other members. During the 2000s the challenges of security and local autonomy in Xinjiang continued, culminating in substantial unrest in July 2009. The central Party-state's response to this, set out in a major work forum on Xinjiang in May 2010, was that development was the key to improving the situation in Xinjiang, and one consequence of this was experiments with cross-border trade and economic zones. Under the twelfth five-year programme (2011–2015) the central government envisaged that Xinjiang should be built into 'an important base for opening to the west'.[27] Plans included (somewhat controversially) turning Kashgar, located in Xinjiang's far west and one of the most ethnic Uyghur cities, into some sort of 'special economic zone',[28] along with Khorgos (Horgos), which was elevated to an administrative city when Xi Jinping visited in 2014, and planned as the starting point of a railway to the Caspian Sea.[29] The focus on Kashgar generated a slogan of 'Shenzhen in the east, Kashgar in the west' and monetary support from the Shenzhen government, but the differences between the two places in their ability to interact with the global economy could not be more stark.

However, whether such development was actually the best way of ameliorating tensions in Xinjiang has been contested, and it has been suggested not only that the ways development has been carried out marginalises the Uyghur population, but that unrest is even to some extent an outgrowth of efforts to engage economically with central and south Asia; in other words, 'state-driven economic development in Xinjiang and Chinese engagement with Central Asia, have not only not helped to mitigate Uyghur discontent with the PRC – they have exacerbated it'.[30] In December 2013 another central work forum appears to have shifted the policy balance back towards 'stability maintenance'.[31]

Section VI of the belt and road vision and action document develops the ideas of Xinjiang opening to its west further, setting out as an aim to 'make good use of Xinjiang's geographic advantage and its role as a window of westward opening-up to deepen communication and cooperation with Central, South and West Asian countries, make it a key transportation, trade, logistics, culture, science and education center, and a core area on the Silk Road Economic Belt.' Since then, some Chinese media reports have promoted Xinjiang as a financial hub for the silk road economic belt.[32] Xinjiang has a key role to play in building the China–Pakistan Economic Corridor, which has been a priority in China's efforts to develop the belt and road. So far this seems to be the main thrust behind designating the region a 'core area' on the silk road economic belt, due to the connectivity from Xinjiang to Pakistan – though the terrain is challenging, and because it traverses territory disputed between India and Pakistan this has fuelled Indian suspicion of the initiative.[33] As of 2015, there does not seem to have been much impact on Xinjiang's trade figures,[34] though the silk road idea was as much about trade going *through* Xinjiang rather than from it; local officials have used the concept of 'corridor economy' (*tongdao jingji*) to reflect the fact that only a small proportion of the region's exports are produced within Xinjiang.[35]

Yunnan, Guangxi, and southeast Asia[36]

The Xinjiang approach displays both similarities and differences to the experiences of Yunnan province, in southwest China. Yunnan has an external land border with Myanmar, Laos and Vietnam, and within China is adjacent to the Tibet

Autonomous Region, Sichuan, Guizhou, and Guangxi. Unlike Xinjiang, whose resource endowments meant that it entered the 1980s a relatively well-off province by Chinese standards (though it has since fallen down the league tables), Yunnan has long been one of most undeveloped parts of the country, with mountainous terrain and poor infrastructure. In the 1980s, conscious of the development of parts of coastal China, policy makers in Yunnan began to consider how they might reorient the province's development strategy to benefit from the 'fruits' of reform and opening up. Finding inspiration in the historical silk roads that ran through the region,[37] they began to promote the building of infrastructure across borders to the south and southwest, and develop greater trade and investment with the province's neighbours.

These ideas gained momentum from 1992, when national opening policies were formally extended to Yunnan and the province became the geographical region through which China would participate in the new Greater Mekong Subregion (GMS) economic cooperation forum, instigated by the Asian Development Bank (ADB) with China and the five countries of the southeast Asian peninsula (Cambodia, Laos, Myanmar, Thailand, Vietnam). Another regional institution, the Bangladesh–China–India–Myanmar (BCIM) forum was promoted from within Yunnan from 1999 and was used by the provincial government to try to develop new links with Bangladesh and eastern parts of India, though it only became a government-to-government mechanism, rather than a dialogue between researchers (with some officials from Yunnan), in 2013. Yunnan policy makers also looked to find ways that the 2001 China–ASEAN free trade agreement could promote the province's international trade and investment and hence its economic development. A key priority in all this activity was developing infrastructure, promoting trade, and from the mid-2000s encouraging Yunnan companies to invest overseas. These ideas coalesced into building Yunnan into a 'bridgehead' to China's west, affirmed by then president Hu Jintao in 2009 and in the 12th five-year programme.[38] An important reason for the centre's advocacy of this strategy was the development of oil and gas pipelines from the Bay of Bengal through Myanmar and into Yunnan, part of China's evolving energy security strategy.[39]

Unlike in Xinjiang, there were minimal political sensitivities for Yunnan in opening up to neighbouring countries. Yunnan's population is one third made up of non-Han ethnic minority groups, but no group dominates, and there is nothing resembling the sort of independence or autonomy movement in Xinjiang. There are security issues raised by transborder interactions, but these relate more to a range of non-traditional security concerns from trafficking in drugs, people and protected species, to the possible spread of infectious diseases such as HIV. These issues have been addressed to some degree through the ADB-sponsored GMS forum, as well as by the provincial and sub-provincial governments in Yunnan. This highlights the tensions between policies to encourage transborder economic development and the cross-border challenges provinces face, as well as the centre's wariness over these challenges.[40]

During this period, Yunnan was not, however, the only provincial entity in China that imagined commercial interactions with southeast Asia as a way of promoting development. In Yunnan's provincial neighbour, the Guangxi Zhuang Autonomous Region, policy makers were also looking at ways of using Guangxi's 'locational advantage' to develop economic ties with southeast Asia – Guangxi has a land border with Vietnam, but also the potential for coastal trade across the South China Sea. In 2005, China's participation in GMS was broadened to include Guangxi alongside Yunnan, and Guangxi's regional capital Nanning was chosen to host the annual China–ASEAN Expo, a new platform for developing trade ties. Guangxi officials have emphasised in particular the potential of the Beibu Gulf (Gulf of Tonkin) development zone to act as a hub for the region to promote China–ASEAN commercial ties. As with Yunnan, ideas of developing economic corridors (a concept favoured by the ADB) feature heavily in Guangxi policy statements.

The relationship between Yunnan and Guangxi in this context is one that can be characterised as 'competitive internationalization',[41] and this is likely to be a wider feature of provincial engagement with the belt and road initiative. Dealing with the competitive element of this involves each province (or region or zone) identifying its own particular strengths in accordance with the idea of comparative advantage. The way this has played out in respect of the belt and road initiative can be seen in the language the vision and actions document uses about Guangxi and Yunnan respectively (in this order):

- We should give full play to the unique advantage of Guangxi Zhuang Autonomous Region as a neighbour of ASEAN countries, speed up the opening-up and development of the Beibu Gulf Economic Zone and the Pearl River-Xijiang [West River] Economic Zone, build an international corridor opening to the ASEAN region, create new strategic anchors for the opening-up and development of the southwest and mid-south regions of China, and form an important gateway connecting the Silk Road Economic Belt and the 21st-Century Maritime Silk Road.
- We should make good use of the geographic advantage of Yunnan Province, advance the construction of an international transport corridor connecting China with neighbouring countries, develop a new highlight of economic cooperation in the Greater Mekong Sub-region, and make the region a pivot of China's opening-up to South and Southeast Asia.

Fujian, Guangdong, and the historical maritime silk road

As well as analysis of the belt and road initiative focusing on the silk road economic belt, there is a growing literature on the twenty-first-century maritime silk road.[42] In economic and commercial terms the maritime route is probably the more important of the two given the levels of China's trade and investment with southeast Asia and the region's economic dynamism, in contrast to central Asia where the Chinese motivation has been dominated by the development since the mid-2000s of energy and resource imports.

However, there is plenty of evidence that – rather as in Xinjiang, Yunnan, and Guangxi – the idea of developing silk road connectivity has been used by academics and policy makers in both Fujian and Guangdong as part of the regional response to China's engagement with globalization from the 1980s and 1990s onwards. The historical understanding on which this was based can be seen in a series of short books published in the 1980s, *The maritime Silk Road collection*, which 'described cultural exchanges, travelers, and famous ports of the Silk Road'.[43] Moreover, these discussions have not simply been initiated from within China, but have been part of wider international projects designed to imagine – or recreate – silk road connectivity in an era of globalization.

This theme is explored in research by historian and sociologist Ana Maria Candela into the role of local actors from Quanzhou (Fujian province) and Guangzhou (Guangdong) in a UNESCO project 'Integral Study of the Silk Roads' (ISSR), which took place from 1988 to 1997. Two of the conferences organized as part of this project took place in Quanzhou and Guangzhou in early 1991 (whereas planned activity in Xinjiang as part of the same project was cut short due to ethnic disturbances there). Quanzhou is a particularly significant location in Chinese thinking about the maritime silk road, often cited as a the starting point of the historical maritime silk road, and a key hub back in Tang dynasty (618–907 CE) connections with Arab traders, who knew the city as Zayton.[44] Speaking at the conference in Quanzhou, then provincial governor Jia Qinglin (who was later promoted to rank fourth nationally in the Communist Party hierarchy) 'pointed out that "China is in essence a Silk Road Country" that has a long history of forging connections with other cultures through its participation in the Silk Road networks', sending a message not just that China was open to the outside world at that point in time, but naturalizing this state by 'embrac[ing] a narrative of Chinese history as having *always* been open to the world'.[45]

This focus on the maritime silk road led to descriptions of China as a water-based or ocean-based civilization, a perspective that was to become controversial after the documentary *River Elegy* was banned in 1988 for being too Western oriented; the documentary had highlighted the maritime theme, suggesting that China's revitalization needed the mixing of the waters of the Yellow River with those of the blue sea. Nonetheless the maritime link continued to resonate in south China, and ideas similar to those discussed in Quanzhou informed thinking in Guangdong. The parallel regional conference in Guangzhou in 1991 differed from its counterpart in Quanzhou in two ways, first by 'locat[ing] the water-based origins of Chinese culture within a Cantonese historical narrative' particular to Guangdong, and second by claiming an 'intercontinental history for China'. But in the context of south China's integration into the global economy, 'South China localities and local officials . . . had strong incentives for supporting further economic reforms and encouraging the active recovery of the region's maritime histories'.[46]

China–Europe trains: marking out the silk road economic belt

Finally, this section returns briefly to discussion of Chongqing as the inspiration for a series of freight train routes developed westwards across the Eurasian continent.[47] These prefigured the shape of the belt and road initiative in a number of ways. The train services from Chongqing to Europe represent precisely the sort of connectivity subsequently advocated under the silk road economic belt, as does the Chongqing government's addition of connectivity to global markets through the southwest into southeast Asia via Kunming and along the Yangzi river to Shanghai and beyond. The language of Chongqing as an 'inland high point of opening up' used in the city from around 2009 has been echoed in the vision and actions document, which envisages such a role for the Chengdu–Chongqing urban cluster. While it is difficult to point to internal policy-making processes that might directly link these earlier policy developments in Chongqing to the development of the belt and road initiative as a national framework, they certainly seem to constitute sub-national policy ideas and practice which informed the national initiative.

In particular, the China–Europe freight train route pioneered by Chongqing was followed by the development of similar train services from cities across China, such as Chengdu, Zhengzhou, and Wuhan. In a plan for the development of China–Europe routes from 2016–2020, the National Development and Reform Commission (NDRC – the main central government ministry for coordinating economic policy) recorded that, at the end of June 2016, there had been 1,881 trains, starting from 16 cities with 12 destination cities in Europe, over 39 lines, transporting US$17 billion of combined exports and imports. A 2020 target was set for 5,000 trains per year across three broad corridors, western (with two main routes), central, and eastern. Other plans include the use of the train routes for international postal services (with routes planned from five Chinese cities, Zhengzhou, Chongqing, Urumqi, Suzhou, and Harbin), and the development of logistics hubs and related facilities to support the freight train services. By March 2017, a substantial number of train services had already been launched to cities in Europe and other places in between China and Europe; these included some routes, such as one to London, which were not included in the NDRC plan (Table 5.1 sets out details of these routes at March 2017).

The attraction of these freight trains from a business perspective is that they lie somewhere between air freight and sea freight in terms of price and speed. At about twice the price of sea freight, rail transportation makes most sense for higher-value goods on more time sensitive orders, and this is reflected in the fact that consumer electronics companies such as HP have been among the first to try the routes. The head of one logistics firm has been quoted as saying that 'our pitch is that rail is 3–5 times cheaper than air, 3–5 times faster than ocean and with 3–5 times better coverage than both'.[48] At the same time, there are numerous challenges and limitations to these routes, from the different gauges used in rail systems in countries across Eurasia, to weather conditions that mean goods can need temperature-controlled containers, and limited imports into China along the routes

TABLE 5.1 China–Europe freight trains (at March 2017)

Origin	Destination	Month inaugurated
Chongqing	Duisburg, Germany	January 2011
Chengdu, Sichuan	Lodz, Poland	December 2012
Zhengzhou, Henan	Hamburg, Germany	July 2013
Suzhou, Jiangsu	Warsaw, Poland	September 2013
Guangzhou, Guangdong	Moscow, Russia	June 2014
Changsha, Hunan	Duisburg, Germany	October 2014
Yiwu, Zhejiang	Madrid, Spain	November 2014
Harbin, Heilongjiang	Hamburg, Germany	June 2015
Kunming, Yunnan	Rotterdam, Netherlands	July 2015
Shenyang, Liaoning	Hamburg, Germany	October 2015
Yiwu, Zhejiang	Tehran, Iran	May 2016
Chengdu, Sichuan	Rotterdam, Netherlands	June 2016
Yiwu, Zhejiang	Mazar-i-Sharif, Afghanistan	August 2016
Nantong, Jiangsu	Hairatan, Afghanistan	August 2016
Tianjin	Minsk, Belarus	November 2016
Yiwu, Zhejiang	London, United Kingdom	January 2017
Chengdu, Sichuan	Minsk, Belarus	March 2017

Source: Compiled by the author from multiple sources

that so far have been used mainly for exports from China.[49] In spite of the number of routes developed, the volumes of trade remain small in relative terms.

Provincial and municipal responses to the belt and road initiative

Looking from a sub-national level is relevant to the future implementation of the belt and road initiative as well as its origins. By setting out brief aspirations for provinces and cities across the country, the section of the vision and actions document on China's regions (Section VI) – which sits in between two sections that deal with implementation through 'cooperation mechanisms' and 'China in action' – should be seen as a key way of exhorting implementation of the initiative at the regional level. This is where the lead for implementation of most economic-related policies in China lies, and as with many national-level policy priorities, a subsequent bureaucratic process of developing provincial policy has resulted in a steady stream of provincial implementation plans for the belt and road being made public. According to one report, in 2015 two thirds of Chinese provinces cited the 'belt and road' as a development priority, included it in their annual work plans and identified particular investment projects that would support it. There is a competitive dynamic at play, with provinces aiming to justify projects or attract investment that can be related to the silk road framework. These specific plans have varied in length, timeline, detail and focus, but reveal a number of common themes about regional responses to the belt and road initiative.[50]

Two Chinese scholars have categorised 17 provincial plans for implementing the silk road economic belt into three types: first those provinces that see the belt as 'the main or material route to opening up and have clear local plans and programs connected with the overall strategy'; second, provincial governments that integrate the belt into 'existing local development plans'; and third, those that connect the belt with 'specific projects and in specific development directions'. The first group consists of Shaanxi, Xinjiang, Gansu and Ningxia, and the second includes Chongqing, Qinghai, Yunnan, Sichuan, Shanxi, Zhejiang, Jiangsu,[51] Shandong, and Hubei.[52] All of the first group and most of the second are provinces in western China (the first four all in the northwest) that are not only relatively underdeveloped compared to national averages, but also have low ratios of foreign trade to GDP (the two significant exceptions are Zhejiang and Jiangsu). Their proactive engagement with the initiative therefore supports the argument that one motivation is to stimulate economic development and openness in these parts of inland China.

Echoing what has been argued above about the regional origins of the belt and road initiative, whatever priority they attach to the belt or road, these provincial implementation plans generally demonstrate a much greater degree of continuity with pre-existing regional or provincial policy than the setting out of new ideas. For example, in central China, the Hunan plan talks about 'upgrading foreign trade', while Henan's emphasises familiar ideas of developing an 'integrated transport hub and commercial logistics centre', and Jiangxi's plan seems to latch onto the initiative as a means of furthering long-standing goals of increasing foreign trade and inward and outward foreign direct investment, for which it identifies specific targets. In the southwest, Guangxi's plan covers eight key points that effectively reiterate existing external development strategy, from developing 'cross-border industry chains' and 'industrial cooperation parks' to 'extending maritime cooperation' and upgrading the China–ASEAN Expo and the Pan-Beibu Gulf (Gulf of Tonkin) Economic Cooperation Forum. Meanwhile, other than in raising the very broad new concept of building a 'golden section' of the silk road economic belt, the plan of the northwest province of Gansu reiterates standard language about developing connectivity and being a gateway for China's westward opening. Other than Gansu's reference to achieving the five connectivities by the middle of the century, the timelines mostly identify goals for 2020, with some adding more general targets for 2025 (even though most interpretations of the belt and road initiative have been that implementation is envisaged over a longer period). As noted above, Shaanxi has a particular status as the start of the historical silk road, as well as political profile from its connection with Xi Jinping and his family, and provincial officials have responded with rhetoric suggesting opening up is 'in full swing' and highlighting the investment in aerospace and other high-tech industries, as well as outward investment such as an oil refinery in Kyrgyzstan built by Shaanxi Coal and Chemical Industry Group, an example of the outward foreign direct investment by Chinese companies, which the initiative looks to encourage.[53]

At the city level, in the southwest, the party leadership of Kunming, provincial capital of Yunnan, issued opinions on 'serving and integrating with' the belt and road in June 2016.[54] The document is a good example of a subnational response that uses the initiative to reiterate and strengthen existing policy goals, rather than set out new material. The most prominent theme is that identified in the discussion above about Yunnan's strategic positioning, namely to act as a hub for China's opening to south and southeast Asia, and to link domestically to the Yangzi River Delta, the Pan-Pearl River Delta, and the Chengdu–Chongqing city cluster, with an aspiration of strengthening its position as a centre in the economic integration of south and southeast Asia with central and western China by 2050. The plan highlights the development of the central Yunnan and Chenggong 'new areas',[55] the city cluster around Kunming (which was identified in the thirteenth five-year programme), and six corridors radiating out from Kunming, each of which have sectoral priorities attached to them. It talks about developing three hubs: an integrated transport hub, an energy hub that will build on the oil and gas pipelines through Myanmar to Kunming, and an information and data hub, all of which reflect the provincial 'bridgehead' strategy. Other plans include encouraging local commercial banks to develop branch networks in south and southeast Asia, cultural exchanges, and developing Kunming further as a host for regional trade exhibitions. In sum, the city's plan brings together a number of long-standing local aims with other national and provincial priorities, and places them under a belt and road umbrella.

To illustrate the subnational responses further, the rest of this chapter returns to look in more detail at one particular example, Fujian, including the city of Quanzhou.

Fujian: a 'core area' of the maritime silk road

The provincial authorities responded to Fujian's status as a 'core area' of the twenty-first-century maritime silk road (MSR) with a detailed policy platform issued in November 2015, entitled the 'Fujian province construction plan for the core area of the 21st century maritime silk road'.[56] As the title suggests, the focus is entirely on the MSR (other than one reference to linking with the silk road economic belt). It reflects a mixture of existing provincial priorities as well as policies stimulated by the initiative. And while the areas that the Fujian government wants to promote under the MSR are consistent with the themes in the vision and actions document, they do not simply map onto its structure. There is little on financial sector connectivity, for example, but much more material on a range of maritime issues than in the vision and actions document. The people-to-people and cultural angles feature prominently throughout. This reflects the notion of comparative advantage in regional policy making, and the basis for the provincial strategy can be seen as reflecting the strengths and characteristics of Fujian: its historical role on the MSR, its connections and proximity to Taiwan, potential and actual maritime links across a wide geography, the number of overseas Chinese and Chinese from Taiwan, Hong Kong and Macao with origins in Fujian, its developed non-state businesses, and maritime economic base.

With this in mind, the policy goals for Fujian can be divided into a number of broad areas. Firstly, there is a strong emphasis on developing logistics infrastructure, from ports along the province's coast to airports and connections inland on rail and road, with several references to facilitating intermodal logistics, and initiatives such as building a MSR logistics information centre. Secondly, the MSR should promote trade and investment in both directions. There are a number of references to supporting outward investment by Fujianese companies across different sectors, as well as to developing 'agricultural production bases' overseas. There is also a desire to use the MSR as a platform to encourage more foreign direct investment in the province, and boost exports to MSR countries, especially of (potential) 'international brands' from Fujian. Other industrial priorities range from light industry sectors to developing e-commerce, and energy cooperation in west Asia. There is space given to developing tourism. A number of trade fairs and other platforms in various cities in Fujian are identified as ways of taking forward the construction of the MSR.

Third, there are injunctions to use the China (Fujian) pilot free trade zone, though these are relatively light on specifics. This was anyway envisaged as a platform for developing international trade and connectivity, so this is another example of an existing area that has been integrated into planning for the implementation of the MSR. Fourth, there are a number of ideas for maritime cooperation. These begin with far-seas fishing, something that has contributed to tensions in the South China Sea over recent years,[57] though the Fujian plan suggests – perhaps rather optimistically – that cooperative relationships in managing fishing can be built across southeast, south, west Asia and Africa. Other areas for maritime cooperation cover environmental protection and research, maritime safety and security (from dealing with natural disasters to anti-terrorism and law enforcement). Finally, there are plenty of references to building cultural and educational linkages, in particular historical exchanges around the MSR, and using Fujian's 'tea culture'.

Even this brief survey highlights extensive Fujianese geographical ambition across the MSR area, and these geographical implications are worth further comment. The plan sets out three broad maritime routes to be developed through Fujian's 'core area' status: south through the South China Sea and Malacca Straits to the Indian Ocean and then extending to the western cooperation corridor with Europe; south through the South China Sea through Indonesia to the southern economic cooperation corridor in the Pacific Ocean; and a northern cooperation corridor to head past South Korea and Japan (building on traditional partnerships in northeast Asia) and extending to the Russian far east and north America. In terms of trade, the plan talks of developing new trade growth points with south Asia, west Asia, and Africa – though this would be from a low base. Sister city relations are envisaged across south Asia, west Asia, east and north Africa, and

Australia and New Zealand. Nonetheless, the weight of the plan lies in building on ties with southeast Asia.

Within Fujian, the plan suggests that Quanzhou should be at the forefront, with important roles for the cities of Fuzhou, Xiamen, and Pingtan. There are also references to developing cooperation with neighbouring provinces in China, connecting with the Yangzi and Pearl River Deltas, and to Fujian serving as an 'important route to the sea' in the opening up of central and western Chinese provinces. This document is also notable for its frequent references to Taiwan. These build on existing policy, under which Fujian has been at the geographical forefront of building economic and commercial links between mainland China and Taiwan. The plan looks to build Fujian-Taiwan commercial cooperation across the region and through MSR countries. This puts a little flesh on the bald statement in the vision and actions document to 'make proper arrangements for the Taiwan region to be part of [the initiative]'.[58]

Reflecting its historical role, and status at the forefront under the provincial plan, the Quanzhou city government issued its own action plan in January 2016.[59] This sets out policy goals to build a 'priority area' (先行区 *xianxingqu*) of the MSR, based on Quanzhou's location as the start of the MSR and the priority given to Quanzhou in building the provincial 'core area'. The document highlights ten action points: reviving the Quanzhou port (as noted above, Quanzhou was one of 15 port cities named in the national-level vision and actions document), trade and investment (inwards and outwards), using the strength from overseas Chinese links, expanding a new 'Arabic corridor', upgrading green and smart manufacturing, financial innovation, building the free trade zone, building a modern maritime city, cultural and tourism cooperation, and nurturing human resources and facilitating exchanges.

It also sets out general goals for 2025 and the 'long-term' around freeing up trade, finance and cultural exchanges, and connecting the city within China and along the MSR. There are some specific targets for 2020 for trade and investment with countries along the MSR: foreign trade with these countries should reach USD 25 billion, accumulated outward foreign direct investment USD 2 billion, and accumulated actual inward foreign direct investment USD 3.5 billion, while 30 percent of GDP will be 'derived from trade in services with countries and regions along the maritime silk road'. For reference, Quanzhou's total trade with all countries in 2014 was just shy of USD 31 billion, while actual FDI in Quanzhou in 2014 reached USD 1.49 billion.[60]

The geographical links envisaged reflect the provincial goals, with a priority focus on southeast and south Asia and the Middle East,[61] a desire to expand ties across central Asia, Europe, northeast Asia and Africa, and with cooperation 'reaching to' Oceania, the American continent and Russia's far east.

Similar to the ideas in the provincial plan, the Quanzhou government looks to play to its strengths from its port to cultural resources, overseas Chinese links, entrepreneurs and non-state sector. These drive six priority areas identified in the action plan:

- Promote the involvement of overseas Chinese in the MSR.[62]
- Encourage cooperation between Quanzhou, Taiwan, Hong Kong, and Macao; here Taiwan seems to be the clear priority, including promoting commercial cooperation under the cross-straits economic cooperation framework agreement (ECFA).
- Internationalize the non-state economy and make manufacturing greener and smarter.
- Opening up and innovation in financial services. This includes references to applying to expand financial opening towards Islamic countries, and attracting Islamic capital to participate in building Quanzhou's MSR priority district.
- Promote international cultural exchange and cooperation and Chinese maritime civilization.
- Develop city-to-city ties, including in port infrastructure and the '21st-century maritime silk road city alliance', whose secretariat will be in Quanzhou.

Conclusion

Looking at the belt and road initiative from the perspective of China's regions suggests that – whatever the other national-level factors may be – the initiative appears as an extension, consolidation and political elevation of pre-existing policy ideas and practice at the sub-national level in China. Rather than being a substantially new policy, it builds on the way that these local ideas and practices have already been raised gradually to the national level over recent years, for example in statements by then Party General Secretary Hu Jintao in 2009 that Yunnan should build a 'bridgehead' to southeast and south Asia, and the material in the 12th and 13th five-year programmes on opening up China's border regions. To the extent that the belt and road initiative builds on these local policy frameworks and practices, the prospects of its implementation along the lines envisaged within China should be enhanced, though the response of other 'countries along the belt and road' will be key.

This regional perspective also highlights the economic and commercial drivers behind the initiative, which can be seen as a way of further 'facilitat[ing] the participation of these localities in the global economy'.[63] This may engender shifts in the geography of external economic and social interactions towards 'countries along the belt and/or road' as indicated in many of the provincial plans – though provincial economic-oriented pragmatism means that this is unlikely to be at the expense of concrete opportunities to engage with other countries, and this factor is likely to push Chinese actors to expand the geographical scope of the initiative. The terms of this participation in the global economy will be in China's favour given its relative economic weight, even at the regional level – the economy of Guangdong province, for example, is larger in aggregate terms than 60 of the other countries originally identified along the belt and road.[64] This is all part of a more proactive Chinese shaping of globalization, as the belt and road and China's regional development together contribute to the development of China in an era of globalization.

Notes

1 This chapter uses 'belt and road initiative' rather than 'one belt, one road' except in citations where the latter term has been used in the original.
2 In March 2017, the Chinese government launched a new official website for the initiative: www.yidaiyilu.gov.cn. For earlier substantial analysis see: Peter Ferdinand. 'Westward Ho: The China dream and "one belt, one road"'. *International Affairs*, 92: 4, 2016; Christopher K. Johnson. *President Xi Jinping's 'Belt and Road' initiative: a practical assessment of the Chinese Communist Party's roadmap for China's global resurgence.* Washington, DC: Center for Strategic and International Studies, 2016, www.csis.org/analysis/president-xi-jinping's-belt-and-road-initiative; Tim Summers. 'China's "New Silk Roads": sub-national regions and networks of global political economy.' *Third World Quarterly*, 37: 9, 2016, pp. 1628–1643; Ye Min. 'China and competing cooperation in Asia–Pacific: TPP, RCEP, and the New Silk Road'. *Asian Security*, 11: 3, 2015; Wang Yong, 'Offensive for defensive: the belt and road initiative and China's new grand strategy'. *The Pacific Review*, 29: 3, 2016. At the time of writing the literature on the belt and road has been growing rapidly, an indicator of the attention paid to the initiative.
3 Ruan Zongze, 'What kind of neighbourhood will China build?' *China International Studies*, 45, 2014, pp. 38–39.
4 Chinese Academy of Social Sciences, cited by Hong Kong Trade Development Council (http://beltandroad.hktdc.com/en/country-profiles/country-profiles.aspx, accessed 29 August 2016). See also: Zhao Hong, 'China's one belt one road: an overview of the debate', ISEAS, 2016, www.iseas.edu.sg/images/pdf/TRS6_16.pdf.
5 *Xinhua.* 'Xi calls for advancing Belt and Road Initiative'. 17 August 2016, http://news.xinhuanet.com/english/2016-08/17/c_135608689.htm.
6 State Council. 'Vision and actions on jointly building belt and road', 28 March 2015, http://news.xinhuanet.com/english/china/2015-03/28/c_134105858.htm.
7 Ferdinand, 'Westward Ho' (see note 2); Johnson, *President Xi Jinping's 'Belt and Road' Initiative* (see note 2).
8 Summers, 'China's "New Silk Roads"' (see note 2).
9 *Xinhua.* 'Zhonghuo renmin gongheguo guomin jingji he shehui fazhan de shisan ge wunian guihua gangyao' (Outline of the 13th five-year programme for national economic and social development of the People's Republic of China), 17 March 2016. http://news.xinhuanet.com/politics/2016lh/2016-03/17/c_1118366322.htm. See Chapter 2 of the present book for background discussion of regional policy during this period.
10 The idea of an 'economic corridor' is to integrate economic activity along a geographical area, in particular through facilitating trade and building infrastructure.
11 Muslim Hui people account for one third of the region's six million population. For references see: Caixin, 'Ningxia official courts investment from Arab Countries', 19 July 2013; China Daily, 'China–Arab states expo opens in Ningxia', 15 September 2013, www.chinadaily.com.cn/china/2013-09/15/content_16971112_4.htm.
12 For more analysis of the position of Tibet, see: Tshering Chonzom Bhutia, 'Tibet and China's "Belt and Road"', 30 August 2016, *The Diplomat*, http://thediplomat.com/2016/08/tibet-and-chinas-belt-and-road/.
13 See Chapters 1 and 2 of the present book.
14 This was the only FTZ to have been approved at the time that the vision and actions document was published in 2015; see Chapter 2 for more background on the FTZs.
15 Liu Junde. 'Regional cooperation in China's administrative region economy: its links with Pan-Pearl River Delta development'. In Xu Jiang and Anthony Yeh, eds. *China's Pan-Pearl River Delta: regional cooperation and development.* Hong Kong: Hong Kong University Press, 2011; Gregory T. Chin. 'The politics of China's western development initiative'. In Ding Lu and William A. W. Neilson, eds. *China's western region development: domestic strategies and global implications.* Singapore: World Scientific, 2004.
16 China Britain Business Council. 'One belt, one road: a role for UK companies in developing China's new initiative', 2015, available through www.cbbc.org/sectors/one-

belt,-one-road/. This report splits these 13 provinces into three regions as follows: northwest (Gansu, Ningxia, Qinghai, Shaanxi, Xinjiang), southwest (Chongqing, Sichuan, Yunnan), south and southeast (Fujian, Guangdong, Hainan, Jiangsu, Zhejiang).
17 CCTV. 'Xi Jinping zai tuijin "yidai yilu" jianshe gongzuo zuotanhui shang qiangdiao, zongjie jingyan jianding xinxin zhashi tuijin, rang "yidai yilu" jianshe zaofu yanxian geguo renmin.' (Xi Jinping remarks at the work forum on "one belt, one road"), 17 August 2016, http://tv.cctv.com/2016/08/17/VIDEv03XNNTONCxVZBZdxeys 160817.shtml.
18 *Xinhua*, 'China focus: Xi's home province a Belt and Road frontier,' 4 March 2016, http://news.xinhuanet.com/english/2016-03/04/c_135155809.htm.
19 This material develops an argument I have made more briefly in Summers, 'China's "New Silk Roads"' (see note 2) and Tim Summers, 'What exactly is "one belt, one road"?' *The World Today*, 71: 5, September 2015, www.chathamhouse.org/publication/twt/what-exactly-one-belt-one-road.
20 Gaye Christoffersen. 'The Russian far east and Heilongjiang in China's Silk Road Economic Belt', 25 April 2016, https://cpianalysis.org/2016/04/25/the-russian-far-east-and-heilongjiang-in-chinas-silk-road-economic-belt/; for background see: Gaye Christoffersen, 'Economic reforms in northeast China: domestic determinants'. *Asian Survey*, 1988, pp. 1245–1263.
21 Gaye Christoffersen. 'Xinjiang and the Great Islamic Circle: the impact of transnational forces on Chinese regional economic planning'. *The China Quarterly*, 133, 1993, p. 136; see also: Michael Dillon, *Xinjiang – China's Muslim far northwest*, Oxford, UK: Routledge, 2004. The idea of an economic circle was 'a regional economic bloc linking a border region with contiguous states for the purpose of trade and economic development', and the three main circles discussed in the 1980s were the Great Islamic Circle, a Great North-east Asian Circle and a South China Economic Circle (Christoffersen, 'Xinjiang', p. 134). The concept of 'great international circles' – which 'may have sounded suspicious to party conservatives because it connoted integration with the capitalist world economy' – fell out of favour after the demise of Zhao Ziyang as Party General Secretary in 1989; Dali Yang. 'China adjusts to the world economy: the political economy of China's coastal development strategy'. *Pacific Affairs*, 64: 1, 1991, p. 50. The language of 'circles' was not used officially in the late 1980s, but does appear briefly in the vision and actions document (Section 3), with slightly different connotations.
22 Michael Clarke, 'Cracks in China's New Silk Road', 15 March 2016, https://cpianalysis.org/2016/03/15/cracks-on-chinas-new-silk-road-xinjiang-one-belt-one-road-and-the-trans-nationalization-of-uyghur-terrorism/.
23 Dillon, *Xinjiang* (see note 21), p. 42.
24 James Millward. 'Positioning Xinjiang in Eurasian and Chinese history: differing visions of the "Silk Road"'. In Michael Clarke and Colin Mackerras, eds. *China, Xinjiang and Central Asia: history, transition and future prospects into the 21st century*. London: Routledge, 2009, pp. 55–74, p. 55, p. 63.
25 Dillon, *Xinjiang* (see note 21).
26 Wang Jianwei. 'China and SCO: Towards a new type of interstate relations'. In Wu Guoguang, ed *China turns to multilateralism: foreign policy and regional security*. London and New York: Routledge, 2007, pp. 104–126.
27 Gangyao. 'Zhonghua Renmin Gongheguo Guomin Jingji he Shehui Fazhan de Shi'er ge Wunian Guihua Gangyao' (Outline of the 12th five-year programme for national economic and social development of the People's Republic of China). Beijing: People's Press, 2011, p. 131.
28 *LA Times*. 'China to turn Silk Road city into special economic zone', 17 November 2010, http://articles.latimes.com/2010/nov/17/world/la-fg-china-kashgar-20101117.
29 *Financial Times*, 'China seeking to revive the Silk Road', 9 May 2016, www.ft.com/content/e99ff7a8-0bd8-11e6-9456-444ab5211a2f.
30 Sean R. Roberts. 'Development with Chinese characteristics in Xinjiang: a solution to ethnic tensions or part of the problem?' In Michael E. Clarke and Douglas Smith, eds.

China's frontier regions: ethnicity, economic integration and foreign relations. London and New York: I. B. Tauris, 2016, p. 24.
31 *Global Times*, 'Xinjiang to see "major strategy shift"', 9 January 2014, originally available at www.globaltimes.cn/content/836495.shtml.
32 *Shanghai Daily*, 'Xinjiang aims for financial hub on economic belt', 8 November 2014, www.shanghaidaily.com/article/article_xinhua.aspx?id=251519.
33 Elizabeth Roche, 'China may invite India to "one belt one road" meet, but Delhi wary', 6 January 2017, www.livemint.com/Politics/iwSTA1EPSjrcnxba0HghTO/China-may-invite-India-to-one-belt-one-road-meet-but-Delh.html. Zhao Hong, 'China's one belt one road: an overview of the debate'. Singapore: ISEAS, 2016. www.iseas.edu.sg/images/pdf/TRS6_16.pdf.
34 Raffaello Pantucci and Anna Sophia Young, 'Xinjiang trade raises doubts over China's "Belt and Road" plan', 10 August 2016, http://blogs.ft.com/beyond-brics/2016/08/10/xinjiang-trade-raises-doubts-over-chinas-belt-and-road-plan/.
35 I am grateful to Weng Xin for this point. This concept of 'corridor economy' is echoed in Yunnan and in other provinces such as Guizhou.
36 This section draws on book-length treatment of these developments in: Tim Summers, *Yunnan: a Chinese bridgehead to Asia: a case study of China's political and economic relations with its neighbours*. Oxford, UK: Chandos, 2013
37 Yang Bin, *Between winds and clouds: the making of Yunnan (second century BCE to twentieth century CE)*. New York: Columbia University Press, 2009.
38 Summers, *Yunnan* (see note 36), pp. 76–78; Gangyao (see note 27), p. 131.
39 Summers, *Yunnan* (see note 36), p. 138.
40 Carla Freeman and Drew Thompson. 'China on the edge: China's border provinces and Chinese security policy'. Center for National Interest and Johns Hopkins SAIS, 2011.
41 Summers, *Yunnan* (see note 36), pp. 103–108 and 179–184.
42 For example, see: Christopher Len. 'China's 21st century maritime silk road initiative, energy security and SLOC access.' *Maritime Affairs*, 11: 1, 2015, pp. 1–18.
43 Ana Maria Candela, '*Qiaoxiang* on the Silk Road: cultural imaginaries as structures of feeling in the making of a global China.' *Critical Asian Studies*, 45: 3, 2013, p. 438.
44 As discussed below, Quanzhou is one of the few cities below the provincial level that has issued its own action plan to follow up the belt and road initiative.
45 Candela, '*Qiaoxiang*' (see note 43), p. 438, emphasis in original.
46 Candela, '*Qiaoxiang*' (see note 43), pp. 439–440.
47 See Chapter 4.
48 Henrik Christensen, chairperson of Silk Route Rail, cited in *Week in China*, 'Belt and road: China's grand gambit', November 2016, www.weekinchina.com/wp-content/uploads/2016/11/WiCFocus13_4Nov2016.pdf, p. 16.
49 Wu, Shang-su. 'The limits of China's "Silk Road" to Europe'. 13 January 2017, *The Diplomat*, http://thediplomat.com/2017/01/the-limits-of-chinas-silk-road-to-europe.
50 Economist Intelligence Unit. 'Prospects and challenges on China's "One Belt, One Road": a risk assessment report', 2015, www.eiu.com/public/topical_report.aspx?campaignid=OneBeltOneRoad, p. 5; Yicai.com, 'Ge di zheng xiandui "yidai yilu" xibu lulu gongcheng you wang shuaixian qidong', 2015, www.yicai.com/news/2015/01/4063211.html.
51 For more detail on Jiangsu's response, see: Li Mingjiang. 'Central–local interactions in foreign affairs'. In John A. Donaldson, ed. *Assessing the balance of power in central–local relations in China*. London and New York: Routledge, 2016, pp. 222–223.
52 Wang Wen and Jia Jinjing (2016). 'Silk Road economic development: vision and path.' IPSI report, p. 113.
53 *Xinhua*. 'China focus: Xi's home province a Belt and Road frontier', 4 March 2016, http://news.xinhuanet.com/english/2016-03/04/c_135155809.htm.
54 Kunming government. 'Kunming shiwei changweihui shenyi tongguo "Kunming fuwu he ronghe 'Yidai yilu' zhanlue de shishi yijian"' ('Meeting of the standing committee of Kunming municipal party committee reviews and approves the implementing opinions

for Kunming to service and integrate into the One Belt, One Road strategy"'), 20 June 2016, http://zfbgt.km.gov.cn/c/2016-06-20/1341698.shtml.
55 See Chapter 2.
56 Fujian government, 'Fujian sheng 21 shiji haishang sichou zhilu hexinqu jianshe fang'an' ('Programme for building Fujian province as a core area of the 21st-century maritime silk road'), 17 November 2015, www.fujian.gov.cn/inc/doc.htm?docid=1095140.
57 Katherine Morton. 'China's ambition in the South China Sea: is a legitimate maritime order possible?' *International Affairs* 92: 4, 2016, pp. 909–940.
58 See the first part of this chapter.
59 Quanzhou government. 'Quanzhou shi jianshe 21 shiji haishang sichou zhilu xianxingqu xingdong fang'an' ('Action plan of Quanzhou city for building the pilot region on the 21st-century maritime silk road), 20 January 2016, www.qzwb.com/qzfb/content/2016-01/20/content_5267417.htm.
60 China cities yearbook press. *Zhongguo chengshi nianjian* (*China cities yearbook*). Beijing: China cities yearbook press, 2015, p. 647.
61 The Quanzhou document uses 'Middle East' (*zhongdong*), while the Fujian provincial document uses 'west Asia' (*xiya*).
62 The text refers to both *huaqiao* (overseas Chinese) and *huaren* (ethnic Chinese).
63 Candela, '*Qiaoxiang*' (see note 43), p. 441.
64 I am grateful to Ben Simpfendorfer for drawing my attention to this point. See also *Week in China*, 'Belt and Road' (see note 48), pp. 11–13.

6
CONCLUSION
China and its regions in an era of globalization

Examining the role of China's sub-national regions in the country's dynamic entanglement with globalization has showed how different regions have played diverse roles both in China's economic and social transformations and in changes in the shape and nature of globalization itself. The initial catalyst for this process was the coincidence of China's 'reform and opening up' policy after 1978 with a new phase of intensified globalization: 'China was ready to enter the world, and the global economy was ready to integrate China',[1] with the result that China's economy embarked on a new development phase at the same time as globalization deepened and broadened.

A number of features of the intensified phase of globalization since the 1980s are particularly relevant when looking at these developments from the perspective of China's regions. A shift from a more managed form of globalization to a neoliberal 'hyperglobalization'[2] from the 1980s saw the rapid development of transnational corporations, facilitated by the policies of many governments to create 'competitive' environments for their operation and attract their investment. This further promoted the development of transnational production networks, into which *parts of China* were gradually incorporated as a result of 'reform and opening up'. China's position in the international division of labour under this period of globalization was generally at a low level, at least up until the global financial crisis. Since then, however, the ongoing shift in the distribution of global political and economic power, the growing strength of Chinese enterprises, the growth in outward foreign direct investment from China, and China's more proactive engagement with global economic governance have given Chinese actors a greater ability to shape globalization.

These actors are not just found in Beijing, in the organs of the central Partystate. Actors in China's regions – governmental, corporate, consumers, and workers – have played an important part in these processes. This has reflected a more general feature of this period of globalization, namely regional formation in response to the flexibilities of late twentieth and early twenty-first century capitalism; during this period, the region – rather than nation-state – has been the 'unit of economic analysis and the territorial sphere most suited to the interaction of political, social and economic processes in an era of "globalization"'.[3] Beyond the formation of regions, the spatial configurations of global political economy have taken the form of networks, into which different parts of China have integrated in different ways and at different points in time. The belt and road initiative looks to develop these structures further, with greater Chinese agency than in the past.[4]

This dynamic global context for China's development, as well as the strategic political and economic concerns of China's leadership, have informed and been informed by regional development. From the 1980s, China's regional policy saw a shift from self-sufficiency and the use of China's inland regions as a defensive front to deal with perceived external threat, to looking to exploit the comparative advantage of different regions in integrating with the global economy. This was followed by a phase where the government sought coordinated and more balanced development across China's regions, while since 2010 new network structures and experimentation with free trade zones and new regional initiatives have characterized regional policy.[5]

Coastal areas, in particular Shenzhen, given its proximity to Hong Kong, were in the forefront of this, primarily through the shift of manufacturing from Hong Kong and the subsequent extension of transnational production networks to incorporate these parts of coastal China. It was this process that led the Pearl River Delta to be described as the 'factory of the world'. But government and businesses in this region have not rested on their laurels, stimulated partly by the need after 2008 to deal with the western-turned-global financial crisis. While some of the older model remains, much of the region has shifted from the labour-intensive and low-value added manufacturing to high-tech sectors and the development of innovative businesses that can push China's position in the global economy (the 'international division of labour') to a higher level. Debates over the extent to which this is possible, and over the role that Chinese companies play in global innovation, will continue; but as a number of examples show – Huawei, Tencent, BGI, DJI, and others – companies based in this part of China are changing the global business landscape, as well as that within China itself. This region is also one of the parts of China – along with regions around Beijing and Shanghai – which is home to many of the companies that are contributing to a more active Chinese shaping of globalization through outward foreign direct investment. The acquisition of a German robotics company in 2016 (which proved controversial in Europe) was by Midea, a typical example of the development of a PRD company from low-end manufacturer to a more influential position in the global economy. The global dynamism and ambition of this region has most recently led the Chinese authorities to plan for the building of a Guangdong-Hong Kong - Macao Greater Bay Area, bringing together the diverse strengths of the economies of this part of China into a US$1.5 trillion economic region.[6]

Meanwhile, the incorporation of parts of coastal China in globalization from the 1980s did little directly to alter the low levels of economic engagement of most of inland China with the global economy. The aspirations of their government officials and business people, though, were high, in particular in the central-western municipality of Chongqing. From the early 2000s there was a gradual 'globalization' of Chongqing's economy, the most prominent features of which have been incorporation into global production networks in the consumer electronics field, followed by the development of ambitious infrastructure linkages from Chongqing across Eurasia to key potential markets in Europe, as well as southwest to southeast Asia, and along the Yangzi river to connect to maritime trading routes. In spite of the tendency to associate disgraced former municipal leader, Bo Xilai, with 'red songs' and a revival of socialism, under his leadership the city did much more to engage with capitalist globalization than reject it, and that process has continued.[7]

The example of Chongqing highlights that domestic integration and connectivity have been growing rapidly, driven by transport and logistics strategies and investment in railway networks, expressways, and civil aviation, as well as information technology. This was also a key finding of my earlier book on the southwest province of Yunnan,[8] which at the same time as developing connectivity to economies across the national borders with Myanmar, Laos, and Vietnam to the southwest, was being integrated much more closely to transport and logistics networks within China. In late 2016, this was manifest in the launch of high-speed rail services from Yunnan's provincial capital, Kunming, to Shanghai, a distance of some 2,660 kilometres. Overall, these developments have tied China's regions more closely to each other and integrated the domestic market in the form of networks between major urban centres, the image developed in the government's own spatial plans from 2010 onwards.[9] Earlier predictions that China's southern or border regions might become more closely tied to external economic relationships than domestic ones have not come to pass.[10]

Nonetheless, it is abundantly clear that developing connectivity across China's land borders has become an increasingly important component of both provincial and national plans. This informs the ideas behind the belt and road initiative, promoted by the Chinese leadership since late 2013. Exploring this from the perspective of China's regions demonstrates both the important regional origins of the initiative, and the central role of subnational actors – particularly provinces and cities – in the initiative's implementation.[11] Along with the examples of the Pearl River Delta and Chongqing, the belt and road initiative also demonstrates the continued dominance of ideas of enhancing global economic, commercial and social interactions – or regional 'openness' in the official language – across China's regions. Indeed, even in Tibet – long excluded from this rhetoric – the goal of opening up has begun to be part of subnational policy and practice, alongside an intensification of linkages within China, whether through developing hubs to serve the domestic market, or in combining external and internal infrastructure development. However, the extent to which this openness is taken on board by provincial policy makers, in theory and in practice, can still vary substantially across the country.[12]

Indeed, although this book has only focused on a few of China's regions, it is readily apparent that the nature of the global interactions of China's regions varies substantially from place to place, and continues to change over time. For what is now being called the Greater Bay Area, the shift has been from incorporation at the lower end of global production networks to becoming a source of innovation, capital, and manufacturing of higher-end products. For Chongqing, the priority has been to cement its place in global production networks in particular sectors, and to launch domestic and global transport projects to overcome the 'tyranny of distance'. Examining the belt and road from a subnational perspective highlights this variation, demonstrating differentiated roles for many provinces and cities, both in the outline plan (the central government's vision and actions document of 2015),[13] and in the subsequent implementation. It also shows that there is a geographical logic to the global interactions of different regions, whereby northeast China's links will look to northeast Asia, while the southeast and southwest will engage more with maritime and peninsular southeast Asia respectively, and the far western region of Xinjiang focuses on connections to central Asia and parts of south Asia, especially through the China–Pakistan Economic Corridor.

This variety in turn reflects a central principle of China's regional policy since the 1980s, namely the idea that regional comparative advantage should drive thinking about the development trajectories of different parts of the country. This concept largely won out over the idea of regional self-sufficiency, which was more prevalent during the first three decades of the PRC, though self-sufficiency has not completely died, and can be seen in elements of provincial and municipal protectionism that still inform the 'regional administrative economy',[14] and even in elements of national policy such as the requirement that provincial governors and city mayors are responsible for the rice basket of their area (though this can be delivered through domestic trade and does not require the same approach to self-sufficiency as in the Mao era).

This variety in regional trajectories and global interactions allows greater resilience across China's economy. It supports the idea that specialization and diversification, and the unique set of capabilities of each region, can enable China as a national whole to dominate at various stages of a fragmented global economy and enhance the overall sustainability of China's economic development.[15] Just as Chongqing and Chengdu have grabbed a significant share of global production in consumer electronics (not as high-tech as it might appear from the local government's statements), Shenzhen is home to globally competitive enterprises such as telecoms company Huawei, and is now the location for most of the global production of newer products such as commercial drones. Provinces or regions of China can adopt 'different positionalities ... in the uneven, regionally variegated pattern by which China has become progressively integrated in global capitalist production networks'.[16]

Global debates about China

These conclusions also shed light on a number of global debates that have intensified with the rise of China. The first concerns the country's future economic trajectories, in particular the ongoing debates about the pace and sustainability of economic growth, and the implications for the global economy. These debates are too often framed in terms of 'China's economy' or an aggregate national GDP growth rate, reducing to a single statistic what is a hugely complex and varied economic area. Sensitivity to the regional picture within China can act as an important corrective to this, and – as argued above – raises the possibility that China's regional diversity could support its ability to remain globally competitive at numerous different levels at the same time.

It also puts into different perspective the scale and impact of China's rise. In 2016, there were three provinces of China – Guangdong, Jiangsu, and Shandong – with aggregate GDP over USD1 trillion (measured in US dollar market exchange rates), larger than the economy of Indonesia, and therefore larger than any southeast Asian economy. Looking at GDP per capita on a purchasing power parity basis gives a better indication of relative levels of development, and on this basis Singapore is well ahead of other regional economies. But here, six Chinese provinces – Tianjin, Beijing, Shanghai, Jiangsu, Zhejiang, Inner Mongolia – come out higher than the next southeast Asian economy, Malaysia.[17] We have reached the point where it is not simply the case that China's economy is dominant across Asia, but where a number of China's provinces are as weighty as some of the major regional economies.

The second major global debate to which this research speaks is over whether China's rise is fundamentally challenging the institutions and practices of global politics and economy, or even upending an existing order (in the language of International Relations scholars, the question is whether China is revisionist or a status quo power). Like the question of China's overall economic trajectory, this is of course a much wider question than can be addressed by the scope of the material in this book. Nonetheless, examining the global interactions of China's regions can contribute meaningfully to the discussion. It shows that at the sub-national level, from provinces downwards – where in practice significant amounts of power in today's China lie – the dominant motivations of political elites are economic and commercial. They are looking to develop their economies, to bring in foreign investment and facilitate their companies' going global, to grasp hold of the best technology and know-how (often still found outside China), and to develop the infrastructure linkages, trade routes (including secure and uninterrupted sea lanes of communication) and wider government and people-based connectivity that facilitates this. At the provincial level, therefore, the predominant interest is in the broad maintenance of the structures and institutions of the existing global political economy and international order, which generally serve China's economic development desires. Provincial voices will ultimately probably not override national security concerns at the top of the Party, but it is worth noting that six of the 25 seats on the Politburo are held by cadres holding positions at the provincial level. Sub-national voices are by definition strong ones in Chinese policy making today, and high-level policy meetings in China – not just on the economically-focused belt and road but also on national security (which in turn requires economic development) – regularly feature presentations from provincial leaders.

The nature of the global interactions of China's regions allows us to develop this point further. Whether it is companies in Shenzhen looking to compete for a position at the commanding heights of the global economy, Chongqing's government looking to locate the city in global production networks and develop infrastructure to allow capital and goods to move more freely, or the uncanny resemblance of the 'belt and road' initiative to a 'spatial fix', which reinforces and reproduces the structures of contemporary global capitalism,[18] China's regions have been integrating proactively into a global capitalist economy. The nature of this engagement is one that reflects and reproduces contemporary capitalist globalization – as Nick Knight puts it in his book on China and globalization, Chinese transnational corporations require the same flexibility as corporations from elsewhere.[19]

This links to the third feature of globalization discussed at the beginning of this book, the spatial structures and reach of the processes of globalization witnessed over the last few decades. Here, the experience of China's regions reinforces two trends. Starting from the point that China's incorporation into globalization from the 1980s was initially an incorporation of *parts of coastal China* into the global economy, the subsequent development of the global interactions of China's regions shows that the further geographical incorporation of China into globalization continued to be uneven and partial. This reflects the very nature of globalization as fragmented – or variegated – rather than homogenous, and as marginalizing territories and populations even as it incorporates new geographies within its reach. More specifically, the spatial structure that emerges is one based around networks and urban clusters. As can be seen in the most recent iterations of regional policy, in the approach from Chongqing, and in the content of the belt and road initiative,[20] such networks have since around 2010 been consciously placed at the heart of regional policy in China. This is not something unique to China, though the explicit mapping of these networks as part of national strategy may be unusual. But these networks are integral to today's global political economy, and further evidence of China's incorporation into its structures.

So far, these concluding comments have been rather 'status quo' in their assessment of the implications of this study for global debates about the impact of the rise of China. But the conclusion should not be that the emergence of China's regions means business as usual. The rise of China – and the concomitant broader shifts in the distribution of global economic weight – has had a transformative impact on the power relations of the global economy, even if the structures have continued to reflect those of contemporary globalization. Evaluating the *extent* to which the scale of China's economic rise has brought it power and influence is hugely debated, and depends crucially on the contested questions of how power and influence are defined.[21] There are strong arguments for considering China's economic transformation to be more of a spread than a rise, and more about quantity than quality, the 'partial power' described by veteran China scholar David Shambaugh. With a relatively small number of exceptions so far, China's ability to break into the 'commanding heights' of the global economy remains limited. And other convincing sceptical accounts of China's rise point to the challenges of human security and development the country still faces or questions of legitimacy.[22]

Nonetheless, in aggregate terms, China's economic development does presage greater weight for the country (and possibly others in the global south) and a relative decline in developed countries' global economic influence. The regional picture outlined in this book – symbolized by the belt and road initiative – also points to the possibility that the global economy will become more Eurasian and less Pacific if China's continental disposition outweighs its maritime one, as it has at various points in the past. There is a new geoeconomics emerging, one in which the balance of voice and power in the system is shifting, even though the overall structures remain those of capitalist globalization. China's regional diversity and scale, and multiple external neighbourhoods, potentially allows it to play multiple roles in these shifts.

China and prospects for globalization

Recent commentary has, however, begun to challenge the paradigmatic status of globalization as a description of our time. Some have suggested that globalization is in retreat, that we are witnessing the beginnings of a period of 'de-globalization', or that protectionism, nationalism, and geopolitics are taking over from economic globalization as the defining feature of global political economy. The evidence cited to support these conclusions is both political and economic, and the politics of Brexit and Trump strengthen the arguments of those who see challenges to globalization. These debates have been picked up in China too.[23]

The economic arguments lead back to the global financial crisis, and are based on the slowdown in global trade volumes and a substantial decline in cross-border capital flows since 2008, as well as the prospect of more regulation of investment and restrictions on migration. These indicators do indeed suggest a slowing, and possibly reverse, of some of the recent features of globalization, though a slowdown does not mean that globalized connectivity will disappear. More persuasive, perhaps, are the arguments that we are witnessing change in the *nature* of globalization – rather than pose a binary question about the demise or otherwise of globalization, the empirical challenge is to work out where change is coming from and what form it will take. For example, at the same time that trade in goods appears to be stagnating, transnational flows of data or information have exploded in scale and look to be more difficult to restrain than flows of goods or even capital.[24]

What does China, and its regions, contribute to this debate? One argument adduced to explain some of the decline in global trade volumes is based on the role of China in global supply chains. As China's economy has grown in scale and strength, parts of manufacturing supply chains that in earlier years were located outside China ('both ends outside') have since been brought within the country's borders. This is precisely the phenomenon of moving 'one end inside' seen in Chongqing's industrial strategy.[25] In other words, China's engagement with contemporary globalization is now being felt as a spread of the structures of a globalized capitalist economy across China rather than simply between the Chinese coast and other economies. This process could continue for some time given the diversity of levels of economic development and factor endowment across China's regions, which will allow different parts of the country to play roles at different levels in the global economy for some years to come. Seen this way, the issue is not so much de-globalization as the further incorporation of more of China into global economy. The answers to current debates about whether this is 'de-globalization' or a globalization that looks different in form and substance will therefore depend at least partly on the approach of China's regions over the coming years.

Policy implications and further research

The implications of this research will vary – as they always do – depending on where the reader sits. In conclusion, I suggest that they fall into three broad areas. First, the geographical scale, diversity and complexity of China's regions requires a step change in the way in which governments, businesses, investors and others outside China engage with the country. It is no longer sufficient to focus attention on Beijing, Shanghai and maybe somewhere in the Pearl River Delta. Even occasional forays inland to Chongqing and Chengdu only begin to scratch the surface of China's diverse and complex development. Trends in GDP growth, investment, and the development of infrastructure suggest shifts in the locations of new economic activity. Many businesses are already engaged with this challenge, thinking in terms of first, second, and third tier cities, or using the concepts of urban clusters as a more manageable and productive approach. But the general debate about China still falls a long way short of grappling with its complexity and diversity. The global interactions – economic, political, and social – of China's regions need to be understood in order to evaluate a changing China.

Second, as was made particularly clear in the discussion of the belt and road initiative, geography and location still matter. It has been convincingly argued that China's rise to date is as more of a regional power than a global one,[26] but what we are now seeing is the effective extension of this regional spread to networks that reach further across Eurasia and into Europe. For those in southeast Asia, dealing with China economically is a familiar challenge but the scale of that is growing faster. For those in Europe China is moving closer, through its sub-national governments and companies, not just at the national level.

Finally, getting to grips with the ongoing development of China's regions and their global interactions creates a substantial research agenda. As noted at the start of this book, the amount of research into China's regions remains small given the scale and diversity of the developments across China. Future research should map in more detail the policy thinking, intra-regional relationships, and global interactions of China's regions.

Notes

1. Andrew J. Nathan and Andrew Scobell. *China's search for security*. New York: Columbia University Press, 2012, p. 245.
2. Dani Rodrik. *The globalization paradox: democracy and the future of the world economy*. New York: W. W. Norton, 2011.
3. Tomaney and Ward, cited in Carolyn Cartier. *Globalizing south China*. Oxford, UK: Blackwell, 2001, p. 264.
4. Tim Summers. 'China's "New Silk Roads": sub-national regions and networks of global political economy'. *Third World Quarterly* 37: 9, 2016, pp. 1628–1643.
5. Outlined in detail in Chapter 2.
6. See Chapter 3.
7. See Chapter 4.
8. Tim Summers. *Yunnan: a Chinese bridgehead to Asia: a case study of China's political and economic relations with its neighbours*. Oxford, UK: Chandos, 2013.
9. See Chapter 2.
10. Gerald Segal. 'The Muddle Kingdom? China's changing shape'. *Foreign Affairs* 73: 3, 1994, pp. 43–58. There may still be some partial exceptions to this, and writing about resource extraction in Tibet, Gabriel Lafitte suggests that it is easier for Chinese enterprises to tap into global supply chains for some minerals than to secure their supply from Tibet; Lafitte. *Spoiling Tibet: China and resource nationalism on the roof of the world*. London: Zed Books, 2013, p. 172.
11. Chapter 5.
12. This point is made in a brief discussion of regional responses to the 'new normal' set out in China's thirteenth five-year programme; see Scott Kennedy and Christopher K. Johnson, 'Perfecting China, Inc.: China's 13th five-year plan', CSIS, 23 May 2016, www.csis.org/analysis/perfecting-china-inc, pp. 44–45.
13. State Council. 'Vision and actions on jointly building belt and road', 28 March 2015, http://news.xinhuanet.com/english/china/2015-03/28/c_134105858.htm.
14. Liu Junde. 'Regional cooperation in China's administrative region economy: its links with Pan-Pearl River Delta development'. In Xu Jiang and Anthony Yeh, eds. *China's Pan-Pearl River Delta: regional cooperation and development*. Hong Kong: Hong Kong University Press, 2011, pp. 63–77.
15. Meg Rithmire. 'China's 'new regionalism': subnational analysis in Chinese political economy'. *World Politics*, 66: 1, 2014, pp. 173–174; Alvin So and Yin-wah Chu. *The global rise of China*. Cambridge, UK: Policy, 2016, p. 132.
16. Andreas Mulvad. 'Competing hegemonic projects within China's variegated capitalism: "liberal" Guangdong vs. "statist" Chongqing'. *New Political Economy*, 20, 2015, p. 208.
17. These comparisons should be taken as indicative only, as they are based on multiple data sources that may not be strictly comparable.
18. David Harvey. *Spaces of capital: towards a critical geography*. Edinburgh, UK: Edinburgh University Press, 2001.
19. Nick Knight. *Imagining globalisation in China: debates on ideology, politics and culture*. Cheltenham, UK: Edward Elgar, 2008, p. 178.
20. Chapters 2, 4, and 5 respectively.
21. For excellent analysis of this question, see: Evelyn Goh. 'Introduction'. In Evelyn Goh, ed. *Rising China's influence in developing Asia*. Oxford, UK: Oxford University Press, 2016.

22 David Shambaugh. *China goes global: the partial power*. Oxford, UK: Oxford University Press, 2013; Mel Gurtov. *Will this be China's century? A skeptic's view*. Boulder, CO and London: Lynne Rienner, 2013.
23 Xu Jian. 'Globalization in reverse and its transformation'. *China International Studies*, July/August 2017, pp. 7–29.
24 James Manyika, Jacques Bughin, Susan Lund, Olivia Nottebohm, David Poulter, Sebastian Jauch, and Sree Ramaswamy. 'Global flows in a digital age'. McKinsey Global Institute, April 2014, www.mckinsey.com/business-functions/strategy-and-corporate-finance/our-insights/global-flows-in-a-digital-age.
25 For details, see Chapter 4.
26 Shambaugh, *China goes global* (see note 22).

BIBLIOGRAPHY

Anderlini, Jamil. *The Bo Xilai scandal: power, death, and politics in China*. London: *The Financial Times*, 2012.
Anderson, Benedict. *Imagined communities: reflections on the origin and spread of nationalism*. London and New York: Verso, 1983.
Ash, Robert F. '"Like fish finding water": economic relations between Hong Kong and China'. In Robert Ash, Peter Ferdinand, Brian Hook, and Robin Porter, eds. *Hong Kong in transition: the handover years*. London: Macmillan, 2000, pp. 58–78.
Ash, Robert, Robin Porter, and Tim Summers. 'Rebalancing towards a sustainable future: China's twelfth five-year programme.' In Kerry Brown, ed. *China and the EU in context: insights for business and investors*. London: Palgrave MacMillan, 2014, pp. 81–141.
Baark, Erik, Yun-Chung Chen, Ngai Pun, and Alvin So. 'Hong Kong and the Pearl River Delta in China: cross-border integration and sustainability'. In Flavia Martinelli, Frank Moulaert, and Andreas Novy, eds. *Urban and regional development trajectories in contemporary capitalism*. London and New York: Routledge, 2013, pp. 127–147.
Barabantseva, Elena V. 'Development as localization: ethnic minorities in China's official discourse on the Western Development Project.' *Critical Asian Studies*, 41: 2, 2009, pp. 225–254.
Bauhinia Foundation. ' "Hun-li" yuanlai buyi, zhonggang huren you fayi?' (' "Marriage and divorce" is not easy, is there a legal basis for mutual recognition between China and Hong Kong?'), 6 August 2016, www.bauhinia.org/index.php/zh-HK/analyses/476.
Bauhinia Foundation Research Centre. 'Creating a World-Class Pearl River Delta Metropolis: Accelerating economic integration between Guangdong and Hong Kong', August 2008, www.bauhinia.org/assets/pdf/research/20081028/PRD%20full%20report-en.pdf.
Beeson, Mark and Fujian Li, *China's regional relations*. Boulder, CO and London: Lynne Rienner, 2014.
Bhattacharya, Abanti. 'Conceptualizing the Silk Road Initiative in China's periphery policy.' *East Asia*, 33: 4, 2016, pp. 309–328.
Bramall, Chris. *In praise of Maoist economic planning: living standards and economic development in Sichuan since 1931*. Oxford, UK: Clarendon Press, 1993.

Brenner, Neil, Bob Jessop, Martin Jones, and Gordon MacLeod, eds. *State/space: a reader*. Malden, MA and Oxford, UK: Blackwell, 2003.

Brødsgaard, Kjeld Erik. *Hainan: state, society, and business in a Chinese province*. New York: Routledge, 2009.

Brown, Kerry. 'Hong Kong ten years on: an assessment of the special administrative region'. Chatham House Asia Programme briefing paper, 2007. www.chathamhouse.org/sites/files/chathamhouse/public/Research/Asia/bp1007hongkong.pdf.

Caixin. 'Chongqing Mayor says rural land reform pilot has been just the ticket'. 17 September 2015, www.caixinglobal.com/2015-09-17/101012190.html.

Caixin. 'Chongqing "xin jingji zhengce"' ('Chongqing's "new economic policies"'), *Zhongguo gaige* (*China Reform*), 325, November 2010, pp. 6–20.

Caixin, 'Ningxia official courts investment from Arab countries', 19 July 2013.

Cameron, Angus and Ronen Palan. *The imagined economies of globalization*. Thousand Oaks, CA: Sage, 2004.

Candela, Ana Maria. '*Qiaoxiang* on the silk road: cultural imaginaries as structures of feeling in the making of a global China.' *Critical Asian Studies*, 45: 3, 2013, pp. 431–458.

Cartier, Carolyn. *Globalizing South China*. Oxford, UK: Blackwell, 2001.

Cartier, Carolyn. 'Origins and evolution of a geographical idea: the macroregion in China.' *Modern China*, 28: 1, 2002, pp. 79–143.

Cartier, Carolyn. 'A political economy of rank: the territorial administrative hierarchy and leadership mobility in urban China'. *Journal of Contemporary China*, 25: 100, 2002, pp. 529–546.

Cartier, Carolyn. 'Uneven development and the time/space economy.' In Wanning Sun and Yingjie Guo, eds. *Unequal China: the political economy and cultural politics of inequality*. New York and Oxford, UK: Routledge, 2013, pp. 77–90.

Castells, Manuel. *The rise of the network society*. Oxford, UK and Malden, MA: Blackwell, 2010.

CCTV. 'Xi Jinping zai tuijin "yidai yilu" jianshe gongzuo zuotanhui shang qiangdiao, zongjie jingyan jianding xinxin zhashi tuijin, rang "yidai yilu" jianshe zaofu yanxian geguo) renmin.' (Xi Jinping remarks at the work forum on "one belt, one road"), 17 August 2016, http://tv.cctv.com/2016/08/17/VIDEv03XNNTONCxVZBZdxeys160817.shtml.

Chan, Anita, Richard Madsen, and Jonathan Unger. *Chen village: revolution to globalization*. Berkeley, CA: University of California Press, 2009.

Chan, Jenny, Ngai Pun, and Mark Selden. 'The politics of global production: Apple, Foxconn and China's new working class'. *New Technology, Work and Employment*, 28: 2, 2013, pp. 100–115.

Chan, Kam Wing. 'China: internal migration.' In Immanuel Ness, ed. *The encyclopedia of global human migration*. Oxford, UK: Blackwell, 2013. http://onlinelibrary.wiley.com/doi/10.1002/9781444351071.wbeghm124/abstract.

Chan, Kam Wing and Will Buckingham. 'Is China abolishing the *hukou* system?' *The China Quarterly*, 195, 2008, pp. 582–606.

Chen Shaofeng. 'Has China's foreign energy quest enhanced its energy security?' *The China Quarterly*, 207, 2011, pp. 600–625.

Chen Shaofeng. 'Yatouhang: Zhongmei Yatai quanshi gengti de fenshuiling?' ('AIIB: A Watershed in Power Transition between US and China?') *The Chinese Journal of American Studies*, 29: 3, 2015, pp. 14–33.

Chen Xiangming. *As borders bend: transnational spaces on the Pacific Rim*. Boulder, CO: Rowman and Littlefield, 2005.

Chen Zhimin, 'Coastal Provinces and China's Foreign policy-making', 2013, www.cewp. fudan.edu.cn/attachments/article/68/Chen%20Zhimin,%20Coastal%20Provinces%20and %20China%27s%20Foreign%20Policy%20Making.pdf.
Cheung, Peter T. Y. and James T. H. Tang. 'The external relations of China's provinces.' In David Lampton, ed. *The making of Chinese foreign and security policy 1978–2000*. Stanford, CA: Stanford University Press, 2001, pp. 91–122.
Chin, Gregory T. 'The politics of China's western development initiative.' In Ding Lu and William A. W. Neilson, eds. *China's western region development: domestic strategies and global implications*. Singapore: World Scientific, 2004, pp. 137–174.
China Britain Business Council. 'One Belt, One Road: A Role for UK Companies in Developing China's New Initiative', 2015, www.cbbc.org/sectors/one-belt,-one-road/.
China Cities Yearbook Press. *Zhongguo chengshi nianjian (China cities yearbook)*. Beijing: China Cities Yearbook Press, 2015.
China.com.cn. 'Chongqing daibiaotuan kaifangri: Bo Xilai Huang Qifan da Zhongwai jizhe tiwen' ('Chongqing delegation's open day: Bo Xilai and Huang Qifan answer questions from Chinese and foreign journalists'), 8 March 2011, www.china.com.cn/ 2011/2011-03/08/content_22085020.htm.
China Daily. 'China-Arab States Expo opens in Ningxia', 15 September 2013, www. chinadaily.com.cn/china/2013-09/15/content_16971112_4.htm.
Chinanews.com. 'Yazhou Zhoukan: Chongqing moshi chuang Zhongguo jingji fangong xin lujing' ('Asiaweek: Chongqing model provides a new path for the Chinese economy'), 6 February 2009, www.chinanews.com/hb/news/2009/02-06/1552353.shtml.
China Statistics Press. *China statistical yearbook 2015*. Beijing: China Statistics Press, 2015.
China Statistics Press. 'Zhongguo diqu jingji jiance baogao' ('Monitoring report on China's regional economy'). Beijing: China Statistics Press, 2010.
Chongqing Customs. 'Huang Qifan tuijie Chongqing shangji: shenchu neilu, que yidian bu bi yanhai cha' ('Huang Qifan promotes Chongqing business opportunities: location in inland China but no worse than the coast'), 25 February 2011, http://chongqing. customs.gov.cn/publish/portal153/tab34627/module76185/info324619.htm.
Chongqing Foreign Trade and Economic Relations Commission. '2014 nian Chongqing shi fuwu maoyi yunxing jiankuang' ('Basic situation of Chongqing's trade in services in 2014'), 2014, www.ft.cq.cn/zxfw/ljcqfwmy/.
Chongqing Liangjiang New Area Management Committee, n.d., 'Chongqing Liangjiang New Area'.
Chongqing News. 'Huang Qifan: "shanzhai qiye" neixuxing jiagong maoyi moshi zhide jiejian' ('Huang Qifan: the domestic demand trade processing model of "copy-cat enter-prises" is worth learning from'), 3 July 2009, http://big5.news.cn/gate/big5/cq. news.cn/2009-07/03/content_16986860.htm.
Christoffersen, Gaye. 'Economic reforms in northeast China: domestic determinants'. *Asian Survey*, 1988, pp. 1245–1263.
Christoffersen, Gaye. 'The Russian Far East and Heilongjiang in China's Silk Road Economic Belt', 25 April 2016, https://cpianalysis.org/2016/04/25/the-russian-far-east-and-heilongjiang-in-chinas-silk-road-economic-belt/.
Christoffersen, Gaye. 'Xinjiang and the Great Islamic Circle: the impact of transnational forces on Chinese regional economic planning.' *The China Quarterly* 133, 1993, pp. 130–151.
Chung, Jao Ho, ed. *Cities in China: recipes for economic development in the reform era*. London and New York: Routledge, 1999.

Chung, Jae Ho, Hongyi Lai, and Jang-Hwan Joo. "Assessing the 'Revive the Northeast' programme." *The China Quarterly*, 197, 2009, pp. 108–125.

Clarke, Michael. 'Cracks in China's New Silk Road', 15 March 2016, https://cpianalysis.org/2016/03/15/cracks-on-chinas-new-silk-road-xinjiang-one-belt-one-road-and-the-transnationalization-of-uyghur-terrorism/.

Clarke, Michael E. and Douglas Smith, eds. *China's frontier regions: ethnicity, economic integration and foreign relations*. London and New York: I. B. Tauris, 2016.

Coates, P. D. *China consuls: British consular officers, 1843–1943*. Oxford, UK: Oxford University Press, 1988.

Constantinescu, Cristina, Aaditya Mattoo and Michele Ruta. 'The global trade slowdown: cyclical or structural?' IMF Working Paper 2015, www.imf.org/external/pubs/ft/wp/2015/wp1506.pdf.

Cui Liru. 'Toward a multipolar pattern: challenges in a transitional stage'. *China-US Focus*, 23 April 2014.

Cui Zhiyuan. 'Partial intimations of the coming whole: the Chongqing experiment in light of the theories of Henry George, James Meade, and Antonio Gramsci.' *Modern China*, 37: 6, 2011, pp. 646–660.

Dent, Christopher. *East Asian regionalism*. London and New York: Routledge, 2008.

DHL. 'Enhanced China-Kazakhstan-CIS-Europe connectivity: DHL and KTZ Express sign MOU to enhance China-Kazakhstan rail freight link', 30 June 2015, www.cn.dhl.com/en/press/releases/releases_2015/local/063015.html.

Dillon, Michael. *Xinjiang: China's Muslim far northwest*. London and New York: Routledge, 2004.

Dirlik, Arif. *Global modernity: modernity in the age of global capitalism*. Boulder, CO: Paradigm Publishers, 2007.

Donaldson, John. *Small works: poverty and economic development in Southwestern China*. Ithaca, NY and London: Cornell University Press, 2011.

The Economist. 'Over troubled water: cross-border transport links are overshadowed by political fears', 13 February 2016, www.economist.com/news/china/21692935-cross-border-transport-links-are-overshadowed-political-fears-over-troubled-water.

The Economist. 'Rich province, poor province.' 1 October 2016, www.economist.com/news/china/21707964-government-struggling-spread-wealth-more-evenly-rich-province-poor-province.

Economist Intelligence Unit. *China hand: the complete guide to doing business in China*, 2014. www.eiu.com.

Economist Intelligence Unit. 'Prospects and Challenges on China's "One Belt, One Road": A Risk Assessment Report', 2015, www.eiu.com/public/topical_report.aspx?campaignid=OneBeltOneRoad.

Enright, Michael. *Developing China: the remarkable impact of foreign direct investment*. London and New York: Routledge, 2016.

European Council on Foreign Relations. '"One Belt, One Road": China's Great Leap Outward', 2015, www.ecfr.eu/page/-/China_analysis_belt_road.pdf.

Fan, Hengshan, ed. *Cujin Zhongbu diqu jueqi: chongda silu yu zhengce yanjiu* (*Promoting the rise of the central regions: major thoughts and policy research*). Beijing: People's Press, 2011.

Ferdinand, Peter. 'Westward ho – the China dream and "one belt, one road"'. *International Affairs*, 92: 4, 2016, pp. 941–958.

Financial Times. 'China seeking to revive the Silk Road', 9 May 2016, www.ft.com/content/e99ff7a8-0bd8-11e6-9456-444ab5211a2f.

Financial Times. 'China seeks to forge foreign demand for its industrial output', 26 January 2015, www.ft.com/content/e8216b60-a152-11e4-8d19-00144feab7de.

Financial Times. 'Chinese overseas lending dominated by One Belt, One Road strategy', 18 June 2015, www.ft.com/content/e9dcd674-15d8-11e5-be54-00144feabdc0.

Financial Times. 'Map: connecting central Asia', 9 May 2016, www.ft.com/cms/s/0/ee5cf40a-15e5-11e6-9d98-00386a18e39d.html.

Financial Times. 'Silk Road revival drives Chinese investment push', 8 August 2016, www.ft.com/content/c2f9f388-5b38-11e6-8d05-4eaa66292c32.

Fitzgerald, John, ed. *Rethinking China's provinces*. London: Routledge, 2002.

Foreign and Commonwealth Office. *The Six-monthly report on Hong Kong, 1 July to 31 December 2003*, 2014. Deposited in Parliament by the Secretary of State for Foreign and Commonwealth Affairs.

Freeman, Carla and Drew Thompson, 'China on the edge: China's border provinces and Chinese security policy'. Center for National Interest and Johns Hopkins SAIS, 2011.

Freund, Caroline and Dario Sidhu, 'Global Competition and the Rise of China', Peterson Institute for International Economics Working Paper, Feb. 2017, https://piie.com/publications/working-papers/global-competition-and-rise-china.

Frieden, Jeffrey. *Global capitalism: its fall and rise in the twentieth century*. New York: W. W. Norton, 2006.

Fujian government. 'Fujian sheng 21 shiji haishang sichou zhilu hexinqu jianshe fang'an' ('Programme for building Fujian province as a core area of the 21st-century maritime silk road'), 17 November 2015, www.fujian.gov.cn/inc/doc.htm?docid=1095140.

Fung Business Intelligence. 'Chongqing: Changjiang shangyou de zhongxin chengshi' ('Chongqing: a central city on the upper reaches of the Yangzi'), 2016, www.fbicgroup.com/sites/default/files/Clusters_Series_special_Chongqing.pdf.

Fung Global Retail and Technology. 'Deep Dive: One Belt One Road – impact on Western multinational companies'. October 2016.

Gangyao. 'Zhonghua Renmin Gongheguo Guomin Jingji he Shehui Fazhan de Shi'er ge Wunian Guihua Gangyao' ('Outline of the 12th five-year programme for national economic and social development of the People's Republic of China'). Beijing: People's Press, 2011.

Garnaud, John. *The rise and fall of the House of Bo*. Melbourne: Penguin, 2012.

Garver, John. 'Development of China's overland transportation links with Central, Southwest and South Asia.' *The China Quarterly*, 185, 2006, pp. 1–22.

Gibson, John and Chao Li. 'Rising regional income inequality in China: fact or artifact?' University of Waikato New Zealand, Department of Economics Working Paper in Economics 09/12, 2012, https://ideas.repec.org/p/wai/econwp/12-09.html.

Giersch, C. Patterson. '"Grieving for Tibet": conceiving the modern state in late-Qing inner Asia'. *China Perspectives*, 3, 2008, pp. 4–19

Global Times. 'Chongqing pioneers China's "socialism 3.0."' 24 September 2010, www.globaltimes.cn/content/576673.shtml.

Global Times. 'Xinjiang to see "major strategy shift"', 9 January 2014. Originally available at www.globaltimes.cn/content/836495.shtml.

Goh, Evelyn, ed. *Rising China's influence in developing Asia*. Oxford, UK: Oxford University Press, 2016.

Goodman, David. 'The campaign to "open up the west": national, provincial-level and local perspectives'. *The China Quarterly*, 178, 2004, pp. 317–334.

Goodman, David. *Centre and province in the People's Republic of China: Sichuan and Guizhou, 1955–1965*. Cambridge, UK: Cambridge University Press, 1986.

Goodman, David, ed. *China's provinces in reform: class, community and political culture*. London and New York: Routledge, 1997.

Goodman, David. 'The politics of regionalism: economic development, conflict and negotiation.' In David Goodman and Gerald Segal, eds. *China deconstructs: politics, trade and regionalism*. London: Routledge, 1994, pp. 1–20.

Goodman, David and Segal, Gerald, eds. *China deconstructs: politics, trade and regionalism*. London: Routledge, 1994

Groenewold, Nicolaas, Anping Chen, and Guoping Lee. *Linkages between China's regions: measurement and policy*. Cheltenham, UK: Edward Elgar, 2008.

Guangdong News. 'Check-in service for Hong Kong International Airport to open in downtown Guangzhou', 3 March 2016, www.newsgd.com/news/2016-03/03/content_143409750.htm.

Guangdong News. 'Dongguan on front lines of robot revolution', 2 March 2016, www.newsgd.com/news/2016-03/02/content_143342870.htm.

Guangdong News. 'HK to bring in private venture capital from SZ', 2 March 2016, www.newsgd.com/news/2016-03/02/content_143324774.htm.

The Guardian. 'The silk railway: freight train from China pulls up in Madrid', 10 December 2014, www.theguardian.com/business/2014/dec/10/silk-railway-freight-train-from-china-pulls-into-madrid.

Gurtov, Mel. *Will this be China's century? A sceptic's view*. Boulder, CO and London: Lynne Rienner, 2013.

Hameiri, Shahar and Lee Jones. 'Rising powers and state transformation: the case of China'. *European Journal of International Relations*, 21: 1, 2016, pp. 72–98.

Han, Enze. *Contestation and adaptation: the politics of national identity in China*. Oxford, UK: Oxford University Press, 2013.

Hansen, Mette Halskov. 'The challenge of Sipsong Panna in the southwest: development, resources and power in a multiethnic China.' In Morris Rossabi, ed. *Governing China's multiethnic frontiers*. Seattle, WA: University of Washington Press, 2004, pp. 53–83.

Hanson, Valerie. *The open empire: a history of China to 1600*. New York and London: W. W. Norton, 2000.

Harrell, Stevan. *Ways of being ethnic in Southwest China*. Seattle, WA: University of Washington Press, 2001.

Harvey, David. *A brief history of neoliberalism*. Oxford, UK: Oxford University Press, 2005.

Harvey, David. *Spaces of capital: towards a critical geography*. Edinburgh, UK: Edinburgh University Press, 2001.

Hendrischke, Hans and Feng Chongyi. *The political economy of China's provinces: comparative and competitive advantage*. London: Routledge, 1999.

Hillman, Ben and Gray Tuttle, eds. *Ethnic conflict and protest in Tibet and Xinjiang*. New York: Columbia University Press, 2016.

Hong, Lijian. 'A tale of two cities: a comparative study of the political and economic development in Chengdu and Chongqing.' In Jao Ho Chung, ed. *Cities in China: recipes for economic development in the reform era*. London and New York: Routledge, 1999, pp. 183–214.

Hong Kong Government. 'CE meets Guangdong officials'. 5 March 2016, www.news.gov.hk/en/categories/admin/html/2016/03/20160305_193711.shtml.

HSBC. 'Asia – Uniquely positioned to capture growth in the Asian century', investor update, 9 June 2015. www.hsbc.com/~/media/hsbc-com/investorrelationsassets/investor-update-2015/asia-china-and-the-pearl-river-delta.

Hu, Angang, Yan Yilong, and Wei Xing. *2030 Zhongguo: maixiang gongtong fuyu [2030 China: towards common prosperity]*. Beijing: People's University Press, 2011.

Huang, Philip C. C. 'Chongqing: equitable development driven by a "Third Hand"?' *Modern China*, 37: 6, 2011, pp. 569–622.

Hung, Ho-fung. *The China Boom: why China will not rule the world*. New York: Columbia University Press, 2016.

Hung, Ho-fung. *China and the transformation of global capitalism*. Baltimore, MA: The John Hopkins University Press, 2009.

InvestHK. *The Greater Pearl River Delta*. Hong Kong: Hong Kong SAR Government, 2014.

Jakobson, Linda and Dean Knox. 'New Foreign Policy Actors in China.' SIPRI Policy Paper 26, September 2010, www.lowyinstitute.org/publications/new-foreign-policy-actors-china.

Johnson, Christopher K. 'President Xi Jinping's "Belt and Road" initiative: a practical assessment of the Chinese Communist Party's roadmap for China's global resurgence'. Washington, DC: Center for Strategic and International Studies, 2016, www.csis.org/analysis/president-xi-jinping's-belt-and-road-initiative.

Kapp, Robert A. *Szechwan and the Chinese Republic: provincial militarism and central power, 1911–38*. New Haven, CT: Yale University Press, 1974.

Kennedy, Scott and Christopher K. Johnson, 'Perfecting China, Inc.: China's 13th five-year plan', CSIS, 23 May 2016, www.csis.org/analysis/perfecting-china-inc.

Knight, Nick. *Imagining globalisation in China: debates on ideology, politics and culture*. Cheltenham, UK: Edward Elgar, 2008.

Kunming government. 'Kunming shiwei changweihui shenyi tongguo "Kunming fuwu he ronghe 'Yidai yilu' zhanlue de shishi yijian"' ('Meeting of the standing committee of Kunming municipal party committee reviews and approves the implementing opinions for Kunming to service and integrate into the "One Belt, One Road strategy"'), 20 June 2016, http://zfbgt.km.gov.cn/c/2016–06–20/1341698.shtml.

Kuznetsov, Alexander. *Theory and practice of paradiplomacy: subnational governments in international affairs*. London and New York: Routledge, 2015.

Lafitte, Gabriel. *Spoiling Tibet: China and resource nationalism on the roof of the world*. London: Zed Books, 2013

Lai, Hongyi. 'Developing central China: a new regional programme'. *China: An International Journal*, 5: 1, 2007, pp. 109–128.

Lardy, Nicholas and Lee Branstetter. 'China's Embrace of Globalization', NBER Working Paper 12373, July 2006, www.nber.org/papers/w12373.pdf.

Larsen, Kjeld Allan. *Regional policy of China, 1949–85*. Manila: Journal of Contemporary Asia Publishers, 1992.

LA Times. 'China to turn Silk Road city into special economic zone', 17 November 2010. http://articles.latimes.com/2010/nov/17/world/la-fg-china-kashgar-20101117.

Leibold, James. 'Ethnic policy in China: is reform inevitable?' *East-West Center policy studies*, 68, 2013, www.eastwestcenter.org/publications/ethnic-policy-in-china-reform-inevitable.

Len, Christopher. 'China's 21st century maritime silk road initiative, energy security and SLOC access.' *Maritime Affairs*, 11: 1, 2015, pp. 1–18.

Li, Keqiang. 'Report on the Work of the Government', 16 March 2017, http://news.xinhuanet.com/english/china/2017-03/16/c_136134017.htm.

Li, Mingjiang, 'Central–local interactions in foreign affairs'. In John A. Donaldson, ed. *Assessing the balance of power in central–local relations in China*. London and New York: Routledge, 2016, pp. 209–228.

Li, Mingjiang. 'From look-west to act-west: Xinjiang's role in China–Central Asian relations.' *Journal of Contemporary China*, 25: 100, 2016, pp. 515–528.

Liu, Junde. 'Regional cooperation in China's administrative region economy: its links with Pan-Pearl River Delta development.' In Xu, Jiang and Anthony Yeh, eds. *China's Pan-Pearl River Delta: regional cooperation and development*. Hong Kong: Hong Kong University Press, 2011, pp. 63–77.

Lin, Justin Yifu and Peilin Liu. 'Development strategies and regional income disparities in China.' In Guanghua Wan, ed. *Inequality and growth in modern China*. Oxford, UK: Oxford University Press, 2008, pp. 56–77.

Lo, Sonny. 'The mainlandization and recolonization of Hong Kong: a triumph of convergence over divergence with mainland China'. In Joseph Cheng, ed. *The Hong Kong special administrative region in its first decade*. Hong Kong: City University of Hong Kong Press, 2007, pp. 179–231.

Mackerras, Colin and Michael Clarke, eds. *China, Xinjiang and Central Asia: history, transition and crossborder interaction into the 21st Century*. London and New York: Routledge, 2009.

Madan, Tanvi. 'Modi's Trip to China: 6 Quick Takeaways.' Brookings Brief, 15 May 2015. www.brookings.edu/research/opinions/2015/05/15-modi-china-takeaways-madan.

Manyika, James, Jacques Bughin, Susan Lund, Olivia Nottebohm, David Poulter, Sebastian Jauch, and Sree Ramaswamy. 'Global flows in a digital age', McKinsey Global Institute, April 2014, www.mckinsey.com/business-functions/strategy-and-corporate-finance/our-insights/global-flows-in-a-digital-age.

Mao, Zedong. 'On the ten great relationships'. Speech given at a meeting of the Political Bureau of the Central Committee of the Chinese Communist Party. 25 April 1956. www.marxists.org/reference/archive/mao/selected-works/volume-5/mswv5_51.htm.

McMahon, Dinny. 'The terrible amusement park that explains Chongqing's economic miracle'. 29 August 2016, *Foreign Policy*, http://foreignpolicy.com/2016/08/29/chongqing-economic-miracle-locajoy-debt-sales-state-owned-enterprises/.

Meisner, Maurice. *Mao's China and after*. New York: Free Press, 1999.

Mercator Institute for China Studies. 'Made in China 2025: the making of a high-tech superpower and consequences for industrial countries', Dec. 2016, www.merics.org/en/merics-analysis/papers-on-china/made-in-china-2025/.

Meyskens, Covell. 'Third Front railroads and industrial modernity in late Maoist China.' *Twentieth-Century China*, 40: 3, 2015, pp. 238–260.

Millward, James. 'Positioning Xinjiang in Eurasian and Chinese history: differing visions of the "Silk Road".' In Michael Clarke and Colin Mackerras, eds. *China, Xinjiang and Central Asia: history, transition and future prospects into the 21st Century*. London: Routledge, 2009, 55–74.

Morton, Katherine. "China's ambition in the South China Sea: is a legitimate maritime order possible?" *International Affairs*, 92: 4, 2016, pp. 909–940

Mosca, Matthew. *From frontier policy to foreign policy: the question of India and the transformation of geopolitics in Qing China*. Stanford, CA: Stanford University Press, 2013.

Mullaney, Thomas. *Coming to terms with the nation: ethnic classification in modern China*. Berkeley, CA: University of California Press, 2011.

Mulvad, Andreas. 'Competing hegemonic projects within China's variegated capitalism: "liberal" Guangdong vs. "statist" Chongqing'. *New Political Economy*, 20, 2015, pp. 199–227.

Nathan, Andrew J. and Andrew Scobell. *China's search for security*. NY: Columbia University Press, 2012.

National Development and Reform Commission. 'Cheng-Yu chengshiqun fazhan guihua.' ('Development programme for the Chengdu–Chongqing urban cluster'), April 2016, www.ndrc.gov.cn/zcfb/zcfbghwb/201605/W020160504587323437573.pdf.

National Development and Reform Commission. 'Outline plan for the reform and development of the Pearl River Delta (2008–2020)'.

National Development and Reform Commission. 'Report on the implementation of the 2015 plan for national economic and social development and on the 2016 draft plan for national economic and social development', delivered at the Fourth Session of the

Twelfth National People's Congress, 5 March 2016, http://news.xinhuanet.com/english/china/2016-03/18/c_135202412_3.htm.

Naughton, Barry. 'Supply-side structural reform at mid-year: compliance, initiative, and unintended consequence.' *China Leadership Monitor*, 51, 2016. www.hoover.org/research/supply-side-structural-reform-mid-year-compliance-initiative-and-unintended-consequences.

Naughton, Barry. "The Third Front: defence industrialization in the Chinese interior." *The China Quarterly*, 115, 1988, pp. 351–386.

NESTA. 'China's absorptive state: innovation and research in China', October 2013, www.nesta.org.uk/publications/chinas-absorptive-state-innovation-and-research-china.

New York Times. 'Hauling new treasure along the Silk Road', 20 July 2013.

Nolan, Peter. *Is China buying the world?* Cambridge, UK and Malden, MA: Polity, 2012.

Nolan, Peter. *Chinese firms, global firms: industrial policy in the age of globalization*. New York: Routledge, 2014.

Oakes, Tim. 'Building a southern dynamo: Guizhou and state power.' *The China Quarterly*, 178, 2004, pp. 467–487.

Oakes, Tim. 'China's provincial identities: reviving regionalism and reinventing "Chineseness".' *The Journal of Asian Studies*, 59: 3, 2000, pp. 667–692.

Panić, Milivoje. *Globalization: a threat to international cooperation and peace?* New York: Palgrave Macmillan, 2011.

Pantucci, Raffaello and Anna Sophia Young. 'Xinjiang trade raises doubts over China's "Belt and Road" plan', 10 August 2016, http://blogs.ft.com/beyond-brics/2016/08/10/xinjiang-trade-raises-doubts-over-chinas-belt-and-road-plan/.

Peck, Graham. *Two kinds of time*. Cambridge, MA: Houghton Mifflin, 1950.

Penna, Michele. 'China's Marshall Plan: all Silk Roads lead to Beijing?' *World Politics Review*, 9 December 2014, www.worldpoliticsreview.com/articles/14618/china-s-marshall-plan-all-silk-roads-lead-to-beijing.

People.cn. 'Xi Zhuxi de "Silu xinyu"' ('President Xi's "new words on the Silk Roads"'). 2 July 2014, http://politics.people.com.cn/n/2014/0702/c1001-25226190.html.

People's Daily. '"Yidai yilu" zhangxian kaifang yu baorong' ('"One belt, one road" clearly indicates openness and inclusivity'), 2 July 2014, http://politics.people.com.cn/n/2014/0702/c1001-25226192.html.

Perdue, Peter. *China marches west: the Qing conquest of Central Eurasia*. Cambridge, MA: Harvard University Press, 2005.

Phillips, Nicola. 'Power and inequality in the global political economy'. *International Affairs*, 93: 2, 2017, pp. 429–444.

Qin, Guangrong. 'Yunnan guoji tongdao jianshe chuyi' ('Initial Thoughts on the Construction of the Yunnan International Transit Route'). In *Yunnan Nianjian (Yearbook)*, 26–31. Kunming: Yunnan Yearbook Press, 2006.

Qiushi. 'Gongjian "Yidai yilu" zhanlue – kaichuang woguo quanfangwei dui wai kaifang xin geju' ('Strategy for building "One belt, one road" – creating a new situation of comprehensive external opening for China'), 28 February 2015, www.qstheory.cn/dukan/qs/2015-02/28/c_1114429328.htm.

Qu, Baozhi. 'Mainland China-Hong Kong economic relations'. In Joseph Cheng, ed. *The Hong Kong special administrative region in its first decade*. Hong Kong: City University of Hong Kong Press, 2007.

Quanzhou government. 'Quanzhou shi jianshe 21 shiji haishang sichou zhilu xianxingqu xingdong fang'an' ('Action plan of Quanzhou city for building the pilot region on the 21st-century maritime silk road), 20 January 2016, www.qzwb.com/qzfb/content/2016-01/20/content_5267417.htm.

Rithmire, Meg E. 'China's 'New Regionalism': subnational analysis in Chinese political economy.' *World Politics*, 66: 1, 2014, pp. 165–194.

Roberts, Sean R. 'Development with Chinese characteristics in Xinjiang: a solution to ethnic tensions or part of the problem?' In Michael E. Clarke and Douglas Smith, eds. *China's frontier regions: ethnicity, economic integration and foreign relations*. London and New York: I. B. Tauris, 2016, pp. 22–55.

Roberts, Sean R. and Kilic Bugra Kanat. 'China's wild west: a cautionary tale of ethnic conflict and development', 15 July 2013, *The Diplomat*, http://thediplomat.com/2013/07/chinas-wild-west/.

Roche, Elizabeth. 'China may invite India to "one belt one road" meet, but Delhi wary'. 6 January 2017, www.livemint.com/Politics/iwSTA1EPSjrcnxba0HghTO/China-may-invite-India-to-one-belt-one-road-meet-but-Delh.html.

Rodrik, Dani. *The globalization paradox: democracy and the future of the world economy*. New York: W. W. Norton, 2011.

Ruan Zongze. 'What kind of neighbourhood will China build?' *China International Studies*, 45, 2014, pp. 26–50.

Schein, Louisa. *Minority rules: the Miao and the feminine in China's cultural politics*. Durham, UK and London: Duke University Press, 2000.

Schram, Stuart. 'Chairman Hua edits Mao's literary heritage: "On the 10 Great Relationships"', *The China Quarterly*, 69, 1977, pp. 126–135.

Segal, Gerald. 'The Muddle Kingdom? China's changing shape.' *Foreign Affairs*, 73: 3, 1994, pp. 43–58.

Selden, Mark. 'Economic nationalism and regionalism in contemporary East Asia.' *The Asia-Pacific Journal*, 10: 43, 2012, p. 2, http://japanfocus.org/-Mark-Selden/3848.

Sen, Tansen. 'Silk Road diplomacy: twists, turns and distorted history', 23 September 2014, Yale Global Online. http://yaleglobal.yale.edu/content/silk-road-diplomacy-–-twists-turns-and-distorted-history.

Shambaugh, David. *China goes global: the partial power*. Oxford, UK: Oxford University Press, 2013.

Shanghai Daily. 'Xinjiang aims for financial hub on economic belt', 8 November 2014, www.shanghaidaily.com/article/article_xinhua.aspx?id=251519.

Shen Dingli. 'China's "One Belt, One Road" strategy is not another Marshall Plan,' 16 March 2015, www.chinausfocus.com/finance-economy/china-advances-its-one-belt-one-road-strategy/.

Shirk, Susan. *The political logic of economic reform in China*. Berkeley, CA: University of California Press, 1993.

Skinner, G. William, ed. *The city in late Imperial China*. Stanford, CA: Stanford University Press, 1977.

So, Alvin Y. '"One country, two systems" and Hong Kong–China national integration: a crisis-transformation perspective'. *Journal of Contemporary Asia*, 41: 1, 2011, pp. 99–116.

So, Alvin and Yin-wah Chu. *The global rise of China*. Cambridge, UK: Polity, 2016.

Sohu.com. 'Huang Qifan: Chongqing jiang chengwei Zhongguo de Ditelv' (Huang Qifan: Chongqing will become China's Detroit'), 19 July 2013, http://business.sohu.com/20130719/n382067427.shtml.

Solinger, Dorothy. *Regional government and political integration in Southwest China 1949–1954*. Berkeley, CA: University of California Press, 1977.

State Council. 'Guojia xinxing chengzhenhua guihua (2014 nian–2020 nian)' ('National programme for a new type of urbanization, 2014–2020'), 17 March 2014, http://politics.people.com.cn/n/2014/0317/c1001-24649809.html.

State Council. Guowuyuan guanyu yinfa quanguo zhuti gongnengqu guihua de tongzhi (Notification of state council plan on national functional regions), 21 December 2010, www.gov.cn/zwgk/2011-06/08/content_1879180.htm.

State Council. 'Guowuyuan guanyu yinfa Zhongguo (Guangdong) ziyou maoyi shixianqu zongti fang'an de tongzhi' (Notification from the State Council about the approval of the overall programme for the China (Guangdong) pilot free trade zone). April 2015, www.gov.cn/zhengce/content/2015-04/20/content_9623.htm.

State Council. 'Vision and actions on jointly building belt and road', 28 March 2015. http://news.xinhuanet.com/english/china/2015-03/28/c_134105858.htm.

STCN. 'Chongqing 2017 nian fuwu maoyi jinchukou zong'e jiang da 300 yi meiyuan' ('In 2017 Chongqing's total trade in services will reach USD 30 billion'), 12 August 2015. www.stcn.com/2015/0812/12406602.shtml.

Steinfeld, Edward. *Playing our game: Why China's rise doesn't threaten the West*. Oxford, UK: Oxford University Press, 2010.

Subacchi, Paola. *The People's Money: how China is building a global currency*. Columbia, NY: Columbia University Press, 2016.

Summers, Tim. 'China's "New Silk Roads": sub-national regions and networks of global political economy.' *Third World Quarterly*, 37: 9, 2016, pp. 1628–1643.

Summers, Tim. 'China's western regions 2020: their national and global implications'. In Kerry Brown, ed. *China 2020: what policy makers need to know about the new rising power in the coming decade*. Oxford, UK: Chandos, 2011, pp. 137–171.

Summers, Tim. 'Chinese multilateralism: diluting the TPP'. In Adrian Hearn and Margaret Myers, eds. *The changing currents of transpacific integration: China, the TPP, and beyond*. Boulder, CO and London: Lynne Rienner, 2017, pp. 77–86.

Summers, Tim. 'Hong Kong: is the handover deal unravelling?' In Kerry Brown, ed. *The critical transition: China's priorities for 2021*. London: The Royal Institute of International Affairs, 2017, www.chathamhouse.org/publication/critical-transition-chinas-priorities-2021.

Summers, Tim. 'Rocking the boat? China's "belt and road" and global order'. In Anoushiravan Ehteshami and Niv Horesh, eds. *China's presence in the Middle East: the implications of the One Belt, One Road initiative*. London and New York: Routledge, 2018, pp. 24–37.

Summers, Tim. 'Thinking inside the box: China and global/regional governance'. *Rising Powers Quarterly*, 1: 1, 2016, pp. 23–31.

Summers, Tim. 'What exactly is "one belt, one road"?' *The World Today*, 71: 5, September 2015.

Summers, Tim. *Yunnan: a Chinese bridgehead to Asia: a case study of China's political and economic relations with its neighbours*. Oxford, UK: Chandos, 2013.

Swaine, Michael. 'Chinese views and commentary on periphery diplomacy.' *China Leadership Monitor* 44, 2014, www.hoover.org/research/chinese-views-and-commentary-periphery-diplomacy.

Szczudlik-Tatar, Justyna. '"One Belt, One Road": mapping China's new diplomatic strategy'. *Bulletin of the Polish Institute of International Affairs*, 67: 799, 2 July 2015, www.pism.pl/files/?id_plik=20062.

Tian, Xiaowen. 'Deng Xiaoping's *Nanxun*: impact on China's regional development'. In John Wong and Zheng Yongnian, eds. *The Nanxun legacy and China's development in the post-Deng era*. Singapore: Singapore University and Press and World Scientific, 2001, pp. 75–92.

Tshering, Chonzom Bhutia. 'Tibet and China's "Belt and Road".' 30 August 2016, *The Diplomat*. http://thediplomat.com/2016/08/tibet-and-chinas-belt-and-road/.

Tzeng, Fuh-wen. 'The political economy of China's coastal development strategy'. *Asian Survey*, 31: 3, 1991, pp. 270–284.

Wang, Hui. *China's new order: society, politics, and economy in transition*, edited by Theodore Huters. Cambridge, MA: Harvard University Press, 2003.

Wang, Jianwei. 'China and SCO: towards a new type of interstate relations'. In Wu Guoguang, ed. *China turns to multilateralism: foreign policy and regional security*. New York and London: Routledge, 2007, pp. 104–126.

Wang, Jisi. 'North, south, east, and west – China is in the "middle": a geostrategic chessboard.' *China International Strategy Review*, 2013, pp. 27–52.

Wang, Shaoguang. 'Changing models of China's policy agenda setting.' *Modern China*, 34: 1, 2008, pp. 56–87.

Wang, Shaoguang. 'Chinese socialism 3.0', translated in Mark Leonard, ed. *China 3.0*. European Council on Foreign Relations, 2012, pp. 60–67, www.ecfr.eu/page/-/ECFR66_CHINA_30_final.pdf.

Wang, Shaoguang. 'The rise of the regions: fiscal reform and the decline of central state capacity in China'. In Andrew G. Walder, ed. *The waning of the Communist State: economic origins of political decline in China and Hungary*. Berkeley, CA: University of California Press, 1995, pp. 87–113.

Wang, Shaoguang and Hu Angang. *The political economy of uneven development: the case of China*. New York: M. E. Sharpe, 1999.

Wang, Wen and Jia Jinjing. 'Silk Road Economic Development: Vision and Path.' IPSI report, 2016.

Wang, Yiwei. 'China's "New Silk Road": a case study in EU-China relations.' In Alessia Amighini and Alex Berkofsky, eds. *Xi's policy gambles: the bumpy road ahead*. Milan: Italian Institute for International Political Studies, 2015.

Wang, Yong. 'Offensive for defensive: the belt and road initiative and China's new grand strategy.' *The Pacific Review*, 29: 3, 2016, pp. 455–463.

Week in China. 'Belt and Road: China's grand gambit'. 2016, www.weekinchina.com/wp-content/uploads/2016/11/WiCFocus13_4Nov2016.pdf.

Wen Wei Pao. 'Huang Qifan zhili Chongqing wu da "qi" zhao' ('Five "rare" moves in Huang Qifan's governance of Chongqing'), 4 August 2015, http://news.wenweipo.com/2015/08/04/IN1508040026.htm.

World Bank. 'China – From poor areas to poor people: China's evolving poverty reduction agenda – an assessment of poverty and inequality in China'. Washington, DC: The World Bank, 2009, http://documents.worldbank.org/curated/en/2009/03/10444409/china-poor-areas-poor-people-chinas-evolving-poverty-reduction-agenda-assessment-poverty-inequality-china-vol-1-2-main-report.

World Bank. 'China 2030: building a modern, harmonious, and creative society'. Washington DC: The World Bank, 2012. http://documents.worldbank.org/curated/en/2013/03/17494829/china-2030-building-modern-harmonious-creative-society.

Wu, Shang-su. 'The limits of China's "Silk Road" to Europe'. 13 January 2017, *The Diplomat*. http://thediplomat.com/2017/01/the-limits-of-chinas-silk-road-to-europe.

Xi, Jinping. 'Jointly shoulder responsibility of our times, promote global growth'. Speech delivered at World Economic Forum Annual Meeting, Davos, 17 Jan. 2017, http://news.xinhuanet.com/english/2017-01/18/c_135991184.htm.

Xinhua. 'Bullet train service starts on China's most challenging railway.' 1 July 2014. http://news.xinhuanet.com/english/china/2014-07/01/c_133453059.htm.

Xinhua. 'China's Belt, Road initiatives not geo-strategic tool: official', 21 March 2015, www.chinadaily.com.cn/business/cdf/2015-03/21/content_19873903.htm.

Xinhua. 'China exclusive: highway corridor links Chongqing to SE Asia.' 4 August 2016. http://news.xinhuanet.com/english/2016-08/04/c_135564313.htm.

Xinhua. 'China focus: Xi's home province a Belt and Road frontier.' 4 March 2016, http://news.xinhuanet.com/english/2016-03/04/c_135155809.htm.

Xinhua. 'Xi calls for advancing Belt and Road Initiative'. 17 August 2016, http://news.xinhuanet.com/english/2016-08/17/c_135608689.htm.

Xinhua. 'Xi envisions Chongqing as int'l logistics hub, stressed development'. 7 January 2016. http://news.xinhuanet.com/english/2016-01/07/c_134984365.htm.

Xinhua. 'Xi Jinping: China to further friendly relations with neighbouring countries', 26 October 2013, http://en.people.cn/90883/8437410.html.

Xinhua. Zhonghuo renmin gongheguo guomin jingji he shehui fazhan di shisan ge wunian guihua gangyao (Outline of the 13th five-year programme for national economic and social development of the People's Republic of China), 17 March 2016. http://news.xinhuanet.com/politics/2016lh/2016-03/17/c_1118366322.htm.

Xu, Jian. 'Globalization in reverse and its transformation'. *China International Studies*, July/August 2017, pp. 7–29.

Xu, Jiang and Anthony Yeh. 'Political economy of regional cooperation in the Pan-Pearl River Delta.' In Yeh, Anthony G. O. and Jiang Xu, eds. *China's Pan-Pearl River Delta: regional cooperation and development*. Hong Kong: Hong Kong University Press, 2011.

Yang, Bin. *Between winds and clouds: the making of Yunnan (second century BCE to twentieth tentury CE)*. New York: Columbia University Press, 2009.

Yang, Dali. *Beyond Beijing: liberalization and the regions in China*. London and New York: Routledge, 1997.

Yang, Dali. 'China adjusts to the world economy: the political economy of China's coastal development strategy.' *Pacific Affairs*, 64: 1, 1991: 42–64.

Yang, Dali. 'Patterns of China's regional development strategy'. *The China Quarterly*, 122, 1990, pp. 230–257.

Yang, Long, 'China's regional development policy'. In John A. Donaldson, ed. *Assessing the balance of power in central-local relations in China*. London and New York: Routledge, 2016, pp. 76–104.

Ye, Min. 'China and competing cooperation in Asia-Pacific: TPP, RCEP, and the New Silk Road.' *Asian Security*, 11: 3, 2015, pp. 206–224.

Yeh, Emily T. and Elizabeth Wharton. 'Going West and going out: discourses, migrants, and models in Chinese development.' *Eurasian Geography and Economics*, 57: 3, 2016, pp. 286–315.

Yeo, Yukyung. 'Between owner and regulator: governing the business of China's telecommunications service industry.' *The China Quarterly*, 200, 2009, pp. 1013–1032

Yeung, Yue-man and Jianfa Shen. 'Cross-boundary integration'. In Yue-man Yeung, ed. *The first decade: the Hong Kong SAR in retrospective and introspective perspective*. Beijing: The Chinese University Press, 2007, pp. 273–295.

Yicai.com. 'Ge di zheng xiandui "yidai yilu" xibu lulu gongcheng you wang shuaixian qidong', 2015, www.yicai.com/news/2015/01/4063211.html.

Yu, LeAnne. *Consumption in China: how China's new consumer ideology is shaping the nation*. Cambridge, UK: Polity, 2014.

Zhang, Weiwei. *The China wave: rise of a civilizational state*. Singapore: World Scientific, 2011

Zhao, Hong. 'China's One Belt One Road: an overview of the debate'. Singapore: ISEAS, 2016. www.iseas.edu.sg/images/pdf/TRS6_16.pdf.

Zheng, Yongnian. *De facto federalism in China: reforms and dynamics of central-local relations*. Hackensack, NJ: World Scientific, 2007.

Zhong, Yang. *Local government and politics in China: challenges from below*. New York: M. E. Sharpe, 2003.
Zweig, David. *Internationalizing China: domestic interests and global linkages*. Ithaca, NY: Cornell University Press, 2002.
163.com. 'Shangwubu buzhang Gao Hucheng: Guowuyuan jueding xin she 7 ge zimao shixian qu' ('Minister of Commerce Gao Hucheng: State Council decision to establish seven new free trade pilot zones'), 31 August 2016, http://c.m.163.com/news/a/BVQP547F000156PO.html.

INDEX

Locators in *italics* refer to figures and those in **bold** to tables.

administrative region economy 15–16, 25, 32, 108
Asian Development Bank (ADB) 90
autonomous regions 15, 19

Bangladesh–China–India–Myanmar forum (BCIM) 90
Beijing 19, 21, 24, 29, 31, 33, 42, 51, 65, 106, 109, 112
Beijing–Tianjin–Hebei regional integration 29, 30, 74
belt and road initiative 2, 9, 42, 45, 52, 78, 83, 99, 106–111; application to China's regions 83–87; and Chongqing 63, 72, 83; provincial and municipal responses 94–99; regional differences 112; regional origins 87–94; regional policy 29–30
Bo Xilai 72, 75–78, 107
Bretton Woods institutions 3–4

Castells, Manuel 5–6
Chengdu 21, 34, 63, 68, 86, 93, **94**, 108
Chengdu–Chongqing urban cluster 30, 73–74, 78, 86, 93, 96
Chiang Kai-shek 12, 20
China: global debates about 109–111; and globalization 6–12, 105–108; overview of book 1–2; policy implications and further research 112–113; prospects for globalization 111–112

China–Europe trains 93–94, **94**
Chongqing 2, 21, 30, 33, 45, 47, 63, **65**, 78; and belt and road initiative 63, 72, 83; challenging globalization 75–78; global positioning 70–73, 107; international economic and industrial strategy 65–69; national policy experiments 69–70; production 66–68, 108; *see also* Chengdu–Chongqing urban cluster; China–Europe trains; infrastructure: Chongqing; political context: Chongqing; Yangzi River economic belt: Chongqing
Closer Economic Partnership Agreement (CEPA) 53–54
coastal regions: belt and road initiative 86, 87–88, 91–92, 96–99; global context 34, 106–107; Mao Zedong 19–20; 28; *see also* foreign direct investment: coastal regions; Greater Bay Area; 'reform and opening up': coastal regions
Cold War 7, 88
common prosperity 76–77
Communist Party *see* People's Republic of China (PRC)
comparative advantage: Hong Kong 55–56; regional policy 22–24, 41, 43, 85, 91, 96, 106, 108
constructivism 11
consumer electronics industry, Chongqing 66–69

Index

coordinated regional development 25–26, 31, 84, 106

de-globalization 2, 111–112
Deng Xiaoping: coastal regions 23–25; common prosperity 76; Pearl River Delta 42–43; 'reform and opening up' 6, 22; southern tour 7, 23–24, 42
'Develop the West' 25–26, 27–28, 32, 33, 65
devolution to provinces 10, 18, 24
diversification of industry 17–18, 108
diversity across China *see* regional differences

economic development: global debates about China 109; Pearl River Delta 47–48
energy policy 28–29, 90
entrepreneurialism 50–51
environmental issues 7, 26–27, 44, 74, 97
ethnic minorities 19, 25, 90

financial crisis *see* global financial crisis
foreign direct investment: belt and road initiative 95, 97; Chongqing 65–66; coastal regions 34; globalization 106; Greater Bay Area 43, 45, 49; and 'reform and opening up' 7, 23
free trade zones (FTZs): Greater Bay Area 52–53; regional policy 30, 32
Fujian 87, 91–92, 96–99

gas pipelines 28–29, 90, 96
GDP growth: global debates about China 109; global financial crisis 46, 46–47; Hong Kong 56; Pearl River Delta 41–42, 49–50, **50**; regional differences 33
geography *see* regional differences
global capitalism 6–7
global cities 5–6
global context: Chongqing's international strategy 65–69; debates about China 109–111; interactions with China's regions 32–34; Pearl River Delta 41–45, 57, 106; of the Third Front 21; *see also* Chongqing: global positioning
global financial crisis: de-globalization 111; regional policy 27–28; *see also* Greater Bay Area: and global financial crisis
globalization: challenging in Chongqing 75–78; and China 6–12, 105–108; China's prospects for 111–112; meaning of 3–4; overview of book 1–2; production 4–5; spatial structures 5–6, 110; *see also* 'reform and opening up': and globalization
global political economy 5–6, 106, 110
government: hierarchical structure 15; international division of labour 4–5; local government system 16, 17; *see also* People's Republic of China (PRC)
Greater Administrative Regions (GARs) 19–20
Greater Bay Area 2, 31, 41, 54, 57–58, **86**, 108; and global financial crisis 45–53; global interactions 108; Hong Kong 53–57; Pearl River Delta 41–45
Guangdong 22, 24, 28, **50**, 51, 58, 109; belt and road initiative 87, 91–92; free trade zone 52–53; global financial crisis 45–48; Pearl River Delta 41–43; trade 34; *see also* Greater Bay Area
Guangxi 25, 72, 85, 89–91, 95
Guangzhou 41, 44–45, 48, 54, 56, 92; *see also* Greater Bay Area

Hainan 22, 84
Harvey, David 6–7
Hebei 29, 30
Hengqin 52, 86
hierarchical structure of government 15, *16*
historical context 11–12
Hong Kong 2, 22, 32, 43, 44–45, 53–57, **56**, 58, 85, 86, 106
household registration (*hukou*) 27, 31, 51, 69
Hu Angang 24
Huang Qifan 72, 76
Huawei 10, 106, 108
Hu Jintao 26, 69, 70, 77, 90, 99
Hu Yaobang 23
hyperglobalization 4, 6–7

inequality 33; *see also* regional differences
infrastructure: Chongqing 70–74; coordinated regional development 25–26; Greater Bay Area 54–55; openness of China 107–108; *see also* China–Europe trains
inland regions: belt and road initiative 86, 87–88; global context 34, 107; Mao Zedong 19–20; *see also* Chongqing; 'reform and opening up': inland regions
international division of labour 4–5, 7, 69, 105–106
investment *see* foreign direct investment

Jiangsu 21, 28, 30, 34, 45, 49, 95, 109

Kunming 29, 72, 93, 96, 107

Li Keqiang 29, 30, 31, 54, 72, 83
local government system 15–16, *16*, 17
location *see* regional differences

macro-regions: belt and road initiative 85–87; meaning of 16–17; regional policy 27–28
major functional zones 26–27
Malacca dilemma 28–29
manufacturing *see* production
Mao Zedong 19–20, 76
maritime silk road 29, 52, 58, 83, 87, 91, 92, 96–99; *see also* belt and road initiative
Ming dynasty 11

National People's Congress (NPC) 31–32
natural resources 26–27
Naughton, Barry 20–21
neoliberalism 6–7
Nepal 85–86
network society 5–6
Ningxia 12, 18, 19, 85, 95

oil pipelines 28–29, 90, 96
one belt, one road *see* belt and road initiative

Pan-Pearl River Delta (PPRD) 45
Pearl River Delta (PRD) 22, 24, 27, 31, 33, **46**, 72, 86, 96, 98, 106, 112; global emergence 41–45, 106; global financial crisis 45–53; Hong Kong 43, 44–45, 53–57, 58
Peck, Graham 63
People's Republic of China (PRC) 2; coordinated regional development 25–26; political structure 15; regional differences 12; regional policy 19–20, 32; structure of 17; *see also* government
policy *see* regional policy
policy implications, this book's 112–113
political context: Chongqing 63–65; global debates about China 109–110; Hong Kong 54–55
power: devolution to provinces 18; global debates about China 109
production: China and globalization 7, 10, 106; Chongqing 66–68, 108; coastal and inland regions 24; international division of labour 4–5, 7; Pearl River Delta 44, 47–48

Qianhai 30, 52, 57, 86
Qing dynasty 11, 16–17, 41, 63
Qingyuan 48–49; *see also* Greater Bay Area
Quanzhou 86, 87, 92, 98

redistribution policy 19–21
'reform and opening up': coastal regions 22–24, 28; and globalization 6–7, 105; inland regions 23–24, 28; international division of labour 7; Pearl River Delta 42–43; regional policy 22–25
regional differences 1–2; administrative region economies 15–16; belt and road initiative origins 87–94; definition of 10–11; diversification vs. specialization 17–18, 108; global interactions 105–108; historical context 11–12; policy implications 112–113; uneven regional development 22–25; urban areas 33
regional policy: 1950s to Third Front 19–21; to 2020 and beyond 29–32; Chongqing 69–70; development of 16–19; 'Develop the West' 25–26, 27–28; global interactions 32–34; new themes from 2010 26–29; *see also* 'reform and opening up': regional policy
research implications, this book 112–113
revisionism, global debates about China 109–111
Rodrik, Dani 4, 9

self-sufficiency 12, 19–21, 22, 106, 108
Shaanxi 30, 33, 64, 85, 87, 95
Shanghai 12, 19, 21, 23, 24, 28, 30, 33, 42, 45, 51, 52, 63, 65, 69, 72, 74, 77, 86, 93, 106, 109, 112; free trade zone 30, 86
Shanghai Cooperation Organization (SCO) 88–89
Shenzhen 12, 22, 23, 28, 30, 43–45, 50–51, **56**, 56–57, 72, 86, 89, 106, 108, 110; *see also* Greater Bay Area
Sichuan 18, 21, 24, 28, 30, 33, 64–65, 73–74, 90, 95
silk road *see* belt and road initiative
Singapore 42, 70, 109
Skinner, George William 16–17
South China Sea 83, 97
southern tour, Deng Xiaoping 7, 23–24
Soviet Union 7, 20, 21; Soviet central Asia 88

spatial structures, globalization 5–6, 110
special economic zones (SEZs): coastal regions 22–23; and free trade zones 30; Pearl River Delta 43; Xinjiang 89
specialization of industry 17–18, 108
state *see* government
state owned enterprises (SOEs) 70
status quo, global debates about China 109–111
Steinfeld, Edward 6
Straw, Jack 53–54

Taiwan 11, 22, 42, 86, 96–97, 98–99
Tang dynasty 11, 92
Technology 6; Chongqing 66–69; Shenzhen 51
the Third Front 16, 20–21, 64, 66
Three Gorges Dam 64, 66
Tiananmen Square tragedy 7
Tianjin 19, 21, 24, 29, 30, 33, 42, 65, 69, 70, 86, **94**, 109
Tibet 18, 19, 28, 74, 85–86, 107
tourism 10, 85, 86, 97, 98
trade: China and the World Trade Organization 7–8, 9–10; Chongqing 66–69; de-globalization 111–112; international division of labour 4–5; Pearl River Delta 42–44; regional differences 34; *see also* free trade zones (FTZs)
trains, China–Europe *see* China–Europe trains
transport *see* belt and road initiative; China–Europe trains; infrastructure

United States 6, 8, 20
urban areas: to 2020 and beyond 31–32; regional policy 27; *see also* Chengdu-Chongqing urban cluster; regional differences: urban areas

wages, Pearl River Delta 47–48
Wang Shaoguang 24, 76
Wang Yang 47, 75, 77
Wen Jiabao 26, 88
World Economic Forum speech, Xi Jinping 9
World Trade Organization, China's joining 7–8, 9–10

Xiamen 55
Xi Jinping 9, 15, 32, 43, 75; belt and road initiative 83–84, 87, 95; Chongqing 69, 71–72, 78; regional policy 29; Xinjiang 89; *see also* World Economic Forum speech, Xi Jinping
Xinjiang 11, 18, 19, 24, 28, 71, 85, 87, 88–89, 90, 92, 95, 108
Xiong'an New Area 29

Yangzi River Delta (YRD) 22, 24, 27, 29, 30, 31, 45, 72, 86, 96, 98
Yangzi River economic belt 16; Chongqing 66, 72, 74, 78
Yunnan 12, 18, 19, 24, 28, 29, 30, 64, 72, 74, 85, 89–91, 96, 99, 107

Zhang Weiwei 1, 34
Zhao Ziyang 23, 64
Zhejiang 28, 30, 33, 34, 42, 45, 49, 67, 86, 95, 109
Zhu Rongji 24, 44
Zhuhai 22, 23, 42, 43, 44, 52, 55